NASA Reference Publication RP-1403

Force Limited Vibration Testing
Monograph

T. D. Scharton

Jet Propulsion Laboratory
California Institute of Technology
Pasadena, CA 91109

May 1997

Acknowledgments

The work described in this paper was carried out by the Jet Propulsion Laboratory (JPL), California Institute of Technology under a contract with the National Aeronautics and Space Administration (NASA). The support of the NASA Headquarters Office of Chief Engineer (Code AE) for the development and documentation of the force limited vibration testing technology described in this monograph is gratefully acknowledged. The many contributions of the personnel in the JPL Dynamics Environments Group and its supervisor, Dennis Kern, are also appreciated. In addition, Robert Bamford, who originated the effective mass concept at JPL in the 1960's must be thanked for his contribution to this monograph on that subject.

FORCE LIMITED VIBRATION TESTING MONOGRAPH

Changes and Errata

Date	Summary of Change	Approved By
9/27/00	Revised equation (7) in Paragraph 3.3.2 Effective Mass	T. Scharton

Table of Contents

List of Figures

List of Tables

1.0 Introduction

The practice of limiting the shaker force in vibration tests was instigated at the NASA Jet Propulsion Laboratory (JPL) in 1990 after the mechanical failure of an aerospace component during a vibration test. Now force limiting is used in almost every major vibration test at JPL and in many vibration tests at NASA Goddard Space Flight Center (GSFC) and at many aerospace contractors. The basic ideas behind force limiting have been in the literature for several decades, but the piezo-electric force transducers necessary to conveniently implement force limiting have been available only in the last decade. In 1993, funding was obtained from the NASA headquarters Office of Chief Engineer to develop and document the technology needed to establish force limited vibration testing as a standard approach available to all NASA centers and aerospace contractors. This monograph is the final report on that effort and discusses the history, theory, and applications of the method in some detail. To facilitate the application of the method, a more concise description of the key aspects of the approach is presented in a complementary guidelines document, which is available separately and is also contained herein as Appendix C.

2.0 History

Many pioneers of aerospace mechanics, as early as the 1950's, recognized that ignoring the low mechanical impedance of lightweight mounting structures, would lead to unrealistically severe vibration tests of aerospace equipment. Unfortunately, their warnings were not heeded until recently, primarily because the tools were not available to conveniently simulate mechanical impedance in vibration tests. Also, more realistic vibration tests are essential in today's faster-better-cheaper environment, which is incompatible with the traditional approaches of high design margins and extensive developmental testing.

Some key background references and their contributions to the force limited vibration technology described in this monograph are summarized in chronological order in this section. These and other supporting references are included in the bibliography of Appendix B.

Blake 1956 [1] describes the problem of overtesting at resonances of the test item which results from the standard practice of enveloping the peaks in the field acceleration spectral data, for both vibration and shock tests. He proposed a complex, conceptual solution in which the impedance of the mounting structure would be simultaneously measured with a small shaker and emulated by the test shaker.

Morrow 1960 [2] warns against ignoring mounting structure impedance in both vibration and shock tests and points out that impedance concepts familiar to electrical engineers are largely unknown to mechanical engineers. He describes exact impedance simulation using force transducers between the shaker and test item, but points out the difficulties in simulating impedance exactly, phase included. (It is still impractical to specify and control the shaker impedance exactly.)

Salter 1964 [3] calls for two test improvements to alleviate overtesting: 1) multi-point control to reduce the impact of fixture resonances and 2) force limiting to account for the vibration absorber effect at test item resonances. He proposes a very simple method of computing the force limit, i.e. the force is limited to 1.5 times the mass times the peak acceleration, i.e. the acceleration specification. His approach, in conjunction with a review of the force data obtained in the system acoustic tests of the Cassini spacecraft, provides the impetus for what in this monograph is called the semi-empirical method of predicting force limits.

Ratz 1966 [4] who was the chief engineer for MB electronics, designs and tests a new shaker equalizer which uses force feedback to simulate the mechanical impedance of the equipment mounting structure (foundation). The last sentence of his paper, "Use of the new equalizer, therefore, can make a dramatic improvement in vibration simulation, and would seem to be the harbinger of a significant advance in the state of the art of vibration testing," looks forward to what is just now becoming a reality.

Heinricks 1967 [5] and **McCaa and Matrullo 1967** [6] in complementary papers describe an analysis and test, respectively, of a lifting body re-entry vehicle using force limiting to notch a random vibration acceleration spectrum. A complete modal model including the effective mass concepts discussed herein, are developed in the analysis. The analysis also includes a comprehensive finite element model (FEM) simulation of the force limited vibration test, in which the test input forces are limited to the structural limit-load criteria. A vibration test of a scale model vehicle is conducted using single-axis impedance-head force transducers to measure the total force input, and the notching is implemented manually based on force data from low level tests.

Painter 1967 [7] conducts an experimental investigation of the sinusoidal vibration testing of aircraft components using both force and acceleration specifications. The interface forces and accelerations between simulated equipment and an aircraft fuselage were measured and enveloped, and these envelopes are used to control shaker sinusoidal vibration tests of the equipment. It is found that the procedure largely eliminated the high levels of overtesting introduced by the conventional approach.

O'Hara 1967 [8] and **Rubin 1967** [9] in two complementary papers, translate and extend the electrical engineering impedance concepts into mechanical engineering terms. O'Hara points out an important distinction between impedance, the ratio of force to an applied velocity, and mobility, the ratio of velocity to an applied force. In the measurement of impedance, for multiple drive points or degrees of freedom, the motion for each degree of freedom other than the one of interest must be blocked, i.e. constrained to have zero motion, which is difficult to realize experimentally. The mobility concept, on the other hand, requires simply that the force at other degrees of freedom be zero, i.e. free boundary conditions. The impedance and mobility **matrices** are reciprocals. In the simple case where the impedance matrice is diagonal (i.e. the ratios of the force to the motion for different degrees of freedom are zero), this distinction does not exist, and the impedance matrix **components** are simply the reciprocal of the corresponding mobility matrix component and equal to what O'Hara calls pseudo-impedance. (In this monograph, an impedance-like quantity, the apparent mass, i.e. the ratio of force to applied acceleration, is most often employed and to avoid complications, which are considerable, the off-diagonal terms of the impedance matrix are typically neglected.) Rubin developed transmission-matrix concepts, which are very useful for coupling systems together and for analyzing vibration isolation.

Murfin 1968 [10] develops the concept of dual control, the first of several at Sandia National Laboratories to contribute to the technology of force limiting, which is one of the key elements of the force limiting approach described in this monograph. He proposes that a force specification be developed and applied in a manner completely analogous to the acceleration specification. The force specification is to be the smoothed envelope of the force peaks in the coupled system response, i.e. the field environment. Recognizing that such field force data are not readily available, he proposes a method of deriving the forces from the product of the acceleration specification and the smoothed apparent mass of the test item. He ignores the mounting structure impedance. This approach is very similar to that used in the semi-empirical method described in this monograph. In the dual control concept, the vibration controller does not give preferential treatment to either the acceleration or the force specification. The force and acceleration control signals are analyzed in narrow bands, and in each band the shaker drive signal is adjusted until one or the other of the specifications is attained. This typically results in acceleration control off resonances and force control, and acceleration notches, at the resonances. Many older vibration test controllers have an extremal or peak control mode in which the largest of several control signals are compared to the acceleration specification, but do not allow separate reference specifications for acceleration and force or response. To use these older controllers for dual control, it is necessary to use a shaping filter to condition the force or response control signal. Most newer vibration test controllers have the capability to implement Murfin's dual control concept, with some minor variations from vender to vender, with their response limiting feature.

Scharton 1969 [11] develops special, multi-modal, vibration test fixtures which had enhanced modal densities and low rigidities, to mechanically simulate the impedance of large flight mounting structures.

Witte 1970 [12] proposes a method of controlling the product of the force and acceleration which is applicable when no information is available on either the test item or the mounting structure mechanical impedance.

Witte and Rodman 1970 [13] and **Hunter and Otts 1972** [14] continue to pursue the calculation of the force specification by multiplying the acceleration specification times the smoothed apparent mass of the test item; like Murfin and their colleagues at Sandia, they ignore the mounting structure impedance. They use simple parametric models to interpret field data and to study the dynamic absorber effect of the payload at resonance, and they develop special methods of smoothing the test item apparent mass.

Wada, Bamford, and Garba 1972 [15] develop a technique for obtaining an equivalent single-degree-of-freedom system (SDFS) for each eigen-vector when the dynamic characteristics of the structure are available in the form of a finite element model (FEM) or as test data. The masses of these SDFS's are called the "effective mass" which may be defined as the mass terms in a modal expansion of the drive point apparent mass of a kinematically supported system. The effective masses are independent of the modal normalization, and the sum of the effective masses of all modes is equal to the total mass. A complementary term is the residual mass, which is defined as the difference between the total mass and the effective masses of all the modes with natural frequencies lower than the frequency of interest. It is a consequence of its definition that the residual mass must be a decreasing function of frequency (Foster's Theorem). The effective mass is very important concept, because it provides a way to quantify, from either FEM's or test data, the mass of a structure as a function of frequency. The effective mass and residual mass are used in this monograph to characterize the mechanical impedance of both the mounting structure (the source) and the test item (the load).

Martini 1983 [16] describes the advent of the piezo-electric, quartz, multi-component force transducer, which is certainly the most important enabling factor in making force limited vibration testing a reality. The development of quartz multi-component force transducers started in 1965 with a Swiss government project to provide a very stiff sensor to measure cutting forces on machine tools, continued in the seventies with the development of biomedical applications, and finally came into its own in the early 80's when the automotive industry began using six component dynamometers for measuring tire and wheel loads. Piezo-electric, quartz transducers offer the following crucial advantages over strain transducers, which have traditionally been used to measure force: 1. extreme rigidity and therefore high resonance frequency, 2. wide dynamic range $\sim 10^4$, 3. large span-to-threshold value $\sim 10^6$, 4. natural resolution into orthogonal components, 4. compactness, 5. wide temperature range - 200C to +200C, 6. low cross-talk, and 7. direct measurement of total force, rather than strain. As seen from the previous references, many of the ideas behind mechanical impedance simulation, and more specifically force limiting, were reported in the literature some thirty years ago. Yet force measurements were seldom made in vibration tests until the 1990's, when quartz multi-component force transducers became readily available, primarily as a result of their development for other markets.

Judkins and Ranaudo 1987 [17] conduct a definitive series of tests quantifying the degree of overtesting in conventional aerospace vibration tests. Their objective is to compare the damage potential of an acoustic test and a conventional random vibration test on a shaker. The test item consists of a three slice mock-up of a RF component weighing approximately 15 lb. and containing three simulated circuit boards. In the acoustic test, the component is mounted on a 0.5" thick honeycomb panel with dimensions of 28.85" by 45.22".The study shows that the shaker resulted in an overtest factor (ratio of shaker to acoustic test results) of 10 to 100 for peak spectral densities and a factor of ten for G rms's. They point out that significant savings in design schedules and component costs will result from reduced vibration test levels which are developed by taking into account the compliance of the mounting structure in the vibration tests of spacecraft components.

Sweitzer 1987 [18] develop a very simple method of correcting for mechanical impedance effects during vibration tests of typical avionics electronic equipment. In essence, the method is to let the test item have a resonance amplification factor of only the square root of Q, rather than Q as it would on a rigid foundation. This is implemented in the test by notching the input acceleration by the same factor, i.e. the square root of Q. This simple method is attractive because it requires no additional instrumentation or knowledge of the actual impedance of either the test item or the mounting structure.

Piersol, White, Wilby, Hipol, and Wilby 1988 [19] conduct a definitive study of the causes and remedies for vibration overtesting in conjunction with Space Shuttle Sidewall mounted components. (It was while working as a consultant on this program that the author became interested in force limited vibration testing.) They compile an extensive biography, contained in Appendix A of their Phase I Report, of literature relating to impedance simulation and the vibration overtesting problem. One aspect of their study is to obtain impedance measurements on the shuttle sidewall and correlate the data with FEM and semi-empirical models. They propose as a force limit the "blocked force", which is the force that the field mounting structure and excitation would deliver to a rigid, infinite impedance, load. Unfortunately, results show that the blocked force is still very conservative for most aerospace applications, so that the overtesting problem is not much alleviated with this particular choice of a limit. A further drawback of the blocked force is that it does not take into account the impedance of the test item, which is readily available in a vibration test incorporating force transducers.

Scharton and Kern 1988 [20] propose a dual control vibration test in which both the interface acceleration and force are measured and controlled. They derive an exact dual control equation which relates the interface acceleration and force to the free acceleration and blocked force. The exact relationship is of little practical value because current vibration test controllers cannot deal with phase, and the exact characteristics of the mounting structure impedance can not be determined. Alternately, they propose an approximate relationship, for dual extremal control, in which the exact relation is replaced by extremal control of the interface acceleration to its specification and the interface force to its specification, as discussed by Murfin.

Scharton, Boatman, and Kern 1989 [21] describe a dual controlled vibration test of aerospace hardware, a camera for the Mars Observer spacecraft, using piezo-electric force transducers to measure and notch the input acceleration in real time. Since the controller would not allow a separate specification for limiting the force, it was necessary to use a shaping filter to convert the force signal into a pseudo-acceleration. One of the lessons learned from this project was that the weight of the fixture above the force transducers should represent a small fraction (less than 10%) of the test item weight. Otherwise above approximately the first mode of the test item, the force signal will be dominated by the force required to vibrate the dead weight of the fixture.

Smallwood 1989 [22] conducts an analytical study of a vibration test method using extremal control of acceleration and force. He finds that the method limited the acceleration input at frequencies where the test item responses tend to be unrealistically large, but that the application of the method is not straightforward and requires some care. He concluded that the revival of test methods using force is appropriate considering the advances in testing technology in the last fifteen years, and that the method reviewed shows real merit and should be investigated further.

Scharton 1990 [23] analyzes dual control of vibration testing using a simple two-degree-of-freedom system. The study indicates that dual controlled vibration testing alleviates overtesting, but that the blocked force is not always appropriate for the force specification. An alternative method is developed for predicting a force limit, based on random vibration parametric results for a coupled oscillator system described in the literature.

Smallwood 1990 [24] establishes a procedure to derive an extremal control vibration test based on acceleration and force which can be applied to a wide variety of test items. This procedure provides a specific, justifiable way to notch the input based on a force limit.

Scharton 1993 [25] describes application of force limited vibration testing to nine JPL flight hardware projects, one of which is the complete TOPEX spacecraft tested at NASA Goddard Space Flight Center. Two of the cases include validation data which show that the force limited vibration test of the components are still conservative compared with the input data obtained from vibration tests and acoustic tests at higher levels of assembly.

Scharton 1994 [26] describes two applications of force limiting: the first to the Wide Field Planetary Camera II for the first Hubble telescope servicing mission, and the second to an instrument on the Cassini spacecraft.

Scharton 1995 [27] devises a method of calculating force limits by evaluating the test item dynamic mass at the coupled system resonance frequencies. Application of the method to a simple and to a complex coupled oscillator system yields non-dimensional analytical results which may be used to calculate limits for future force limited vibration tests. The analysis for the simple system provides an exact, closed form result for the peak force of the coupled system and for the notch depth in the vibration test. For example, using the simple system results with Q=50 and equal impedance of the flight mounting structure and test item, the input acceleration will be notched by a factor of 31.25 relative to a conventional test. The analysis for the complex system provides parametric results which contain both the effective modal and residual masses of the test item and mounting structure and is therefore well suited for use with FEM models.

Scharton and Chang 1997 [28] describe the force limited vibration test of the Cassini spacecraft conducted in November of 1996. Over a hundred acceleration responses were monitored in the spacecraft vibration test, but only the total axial force is used in the control loop to notch the input acceleration. The force limit specified in the spacecraft vibration test plan is used in the test without any modifications, and many of the major equipment items on the spacecraft reached their flight limit load. The force limit for the complete spacecraft vibration test, as well as the limits for many of the Cassini instrument vibration tests, are developed using a simple, semi-empirical method which requires only the acceleration specification and data from a low level pre-test to determine the apparent mass of the test item. This semi-empirical method of predicting force limits is validated for the instrument tests by comparisons with two-degree-of-freedom analytical models and with interface force data measured at the instrument/spacecraft interface in acoustic tests of the Cassini spacecraft DTM structure. The instrument force limits derived with the semi-empirical method are generally equal to or less than those derived with the two-degree-of-freedom method, but are still conservative with respect to the interface force data measured in the acoustic test.

3.0 Structural Impedance

3.1 The Vibration Overtesting Problem

There are historically three solutions to the vibration overtesting problem: 1) "build it like a brick", 2) mechanical impedance simulation, and 3) response limiting.

Some aerospace components are still "built like a brick" and therefore can survive vibration overtesting and perhaps even an iterative test failure, rework, and retesting scenario. In a few cases, this may even be the cheapest way to go, but the frequency of such cases is certainly much less than it used to be. The two historical methods of alleviating overtesting, impedance simulation and response limiting, are both closely related to force limiting.

3.1.1 Impedance Simulation

In the 1960's, personnel at NASA's Marshall Space Flight Center (MSFC) developed a mechanical impedance simulation technique called the "N plus one structure" concept, which involved incorporating a portion of the mounting structure into the vibration test. A common example would be the vibration test of an electronic board, mounted in a black box. In addition, acoustic tests were often conducted with the test items attached to a flight-like mounting structure. In all these approaches where a portion of the mounting structure, or simulated mounting structure, is used as the vibration test fixture, it is preferred that the acceleration input be specified and monitored internally at the interface between the mounting structure and the test item. If instead, the acceleration is specified externally at the interface between the shaker and mounting fixture, the impedance simulation benefit is greatly decreased. (When the input is defined internally, the "N plus one structure" approach is similar to the response limiting approach, discussed in the next section.)

A second example of mechanical impedance simulation is the multi-modal vibration test fixture [11] which was designed to have many vibration modes to emulate a large flight mounting structure. This novel approach was used in one government program, a Mariner spacecraft, but fell by the wayside along with other mechanical impedance simulation approaches as being too specialized and too expensive. In addition, the concept went against the conventional wisdom of making fixtures as rigid as possible to avoid resonances.

Mechanical impedance simulation approaches are seldom employed because they require additional hardware and therefore added expense. Two exceptions which may find acceptance in this new low cost environment are: 1) deferring component testing until higher levels of assembly, e.g. the system test, when more of the mounting structure is automatically present, and 2) replacement of equipment random vibration tests with acoustic tests of the equipment mounted on a flight-like plate, e.g. honeycomb.

3.1.2 Response Limiting

Most institutions have in the past resorted to some form of response limiting as a means of alleviating vibration overtesting. Response limiting is analogous to force limiting, but generally more complicated and dependent on analysis. Response limiting was used for several decades at JPL but has now been largely replaced by force limiting. In both response and force limiting, the approach is to predict the in-flight response (force) at one or more critical locations on the item to be tested, and then to measure that response (force) in the vibration test and to reduce, or notch, the acceleration input in the test at particular frequencies, so as to keep the measured response (force) equal to or below that limit.

In the case of response limiting, the in-flight response is usually predicted with FEMs. This means that the prediction of response requires an FEM model of the test item and the directly excited supporting structure and the same model is typically used to design and analyze the loads in the test item. In this case the role of the test as an independent verification of the design and analysis is severely compromised. In addition, the model has to be very detailed in order to predict the in-flight response at critical locations, so the accuracy of the predictions is usually suspect, particularly at the higher frequencies of random vibration. By contrast, the interface force between the support structure and test item can be predicted with more confidence, and depends less or not at all on the FEM of the test item.

In addition, it is often complicated or impossible to measure the responses at critical locations on the vibration test item. Sometimes the critical locations are not accessible, as in the case of optical and cold components. In the case of large test items, there may be many response locations of interest; hundreds of response locations may be measured in a typical spacecraft test. For this reason, some institutions rely completely on analysis to predict the responses in flight and in the test, and then a priori shape the input acceleration for the test in order to equate the flight and test responses. Since the uncertainty in the predictions of the resonance frequencies of the item on the shaker is typically 10 to 20 %, any notches based on pre-test analysis must be very wide, and may result in undertesting at frequencies other than at resonances.

There is one form of response limiting which is conceptually identical to force limiting, i.e. limiting the acceleration of the center-of-gravity (CG) or mass centroid of the test item. By Newton's second law, the acceleration of the CG is equal to the external force applied to the body, divided by the total mass. It is indeed much easier to predict the in flight responses of the test item CG, than the responses at other locations. The CG response is typically predicted with FEM's using only a lumped mass to represent the test item. At JPL a semi-empirical curve, called the mass-acceleration curve is usually developed early in the design process to predict the CG response of payloads. Also, any method used to predict the in-flight interface force obviously predicts the CG acceleration as well.

The problem with CG response limiting in the past has been that it is difficult or impossible to measure the acceleration of the CG with accelerometers in a vibration test. Sometimes the CG is inaccessible, or there is no physical structure at the CG location on which to mount an accelerometer. However, there is a more serious problem. The CG is only fixed relative to the structure, when the structure is a rigid body. Once resonances and deformations occur, it is impossible to measure the CG acceleration with an accelerometer. Furthermore attempts to measure the CG response usually overestimate the CG response at resonances, so limiting based on these measurements will result in an undertest. However, the CG acceleration is uniquely determined by dividing an interface force measurement by the total mass of the test item; this technique is very useful and will be discussed subsequently in conjunction with quasi-static design load verification.

3.1.3 Enveloping Tradition

The primary cause of vibration overtesting is associated with the traditional, and necessary, practice of enveloping acceleration spectra to generate a vibration test specification. In the past the overtesting, or conservatism as some preferred to call it, was typically attributed to the amount of margin that was used to envelope the spectral data or predictions. Now it is understood that the major component of overtesting is inherent in the enveloping process itself, and is not within the control of the person doing the enveloping.

In Figure 1, consider the data taken during the TOPEX spacecraft acoustic test [29]. Each of the six curves is a measurement near the attachment point of a different electronic box to a honeycomb panel. The flat trace is the test specification for the random vibration tests conducted on the electronic boxes, a year or so prior to the spacecraft acoustic test. Ideally the specification would just envelope the data, and the agreement is pretty good in the mid-frequency range from 100 to 500 Hz. One might rationalize that below 100 Hz, the random vibration specification is high to account for low frequency transients not simulated in the acoustic test, and above approximately 500 Hz the specification is high to account for direct acoustic excitation of the boxes which is not reflected in the attachment foot data. Therefore one may conclude that the random vibration tests of the TOPEX spacecraft electronic boxes was not unduly conservative, but this would be erroneous. Each of the six curves in Fig. 1 has peaks and valleys, at different frequencies. The specification does a good job of enveloping all the peaks as it should, but what about the valleys. Clearly, the valleys are far below the specification, as illustrated by the dark highlighted curve. The next section, on the dynamic absorber effect, will show that that the frequencies associated with the valleys are very special, in that they represent the resonance frequencies of the boxes with fixed bases, i.e. as they are mounted in the random vibration tests on the shaker. In other words, the random vibration tests resulted in an overtest at the box resonances, by the amount that the valleys in Fig. 1 are below the specification, i.e. typically 10 to 20 dB!

Based on the preceding data, one might argue that random vibration specifications should envelope the valleys, not the peaks of the field data. However, this is not possible and would result in undertesting off the resonances. The best approach, and the one implemented with force limiting, is to retain the traditional vibration test specification, which is the envelope of the peaks, but to notch the input at the resonance frequencies on the shaker to emulate the valleys in the field environment.

3.1.4 Dynamic Absorber Effect

The dynamic absorber effect [30] may be explained with the assistance of Fig. 2, which shows a simple vibratory system consisting of two oscillators, connected in series. The primary oscillator is directly excited and the secondary oscillator is undamped and excited only by virtue of its connection to the first oscillator. The dynamic absorber effect refers to the fact that the motion of the mass of the primary, directly excited, oscillator will be zero at the natural frequency of the secondary oscillator. This statement is true even when the natural frequencies of the two oscillators are different and even when the mass of the secondary oscillator is much less than that of the secondary oscillator is small rather than zero, the motion of the primary mass is small rather than zero.

To apply the dynamic absorber effect to the aerospace vibration testing problem, assume that the two oscillator system in Fig. 2 represents a vibration mode of a flight support structure coupled to a vibration mode of a vibration test item. For example, the support structure might be a spacecraft, and the test item an instrument mounted on the spacecraft. Consider the numerical example illustrated in Fig. 3, for the case where the two uncoupled oscillator natural frequencies are identical, the masses are unity, the base acceleration is unity, and the Q is 50. The ordinates in Fig. 3 are FRF magnitudes, and for convenience the results are discussed in terms of a sinusoidal input. The abscissa in Fig. 3 is frequency, normalized by the natural frequency of an uncoupled oscillator. Fig. 3a is the magnitude of the coupled system interface

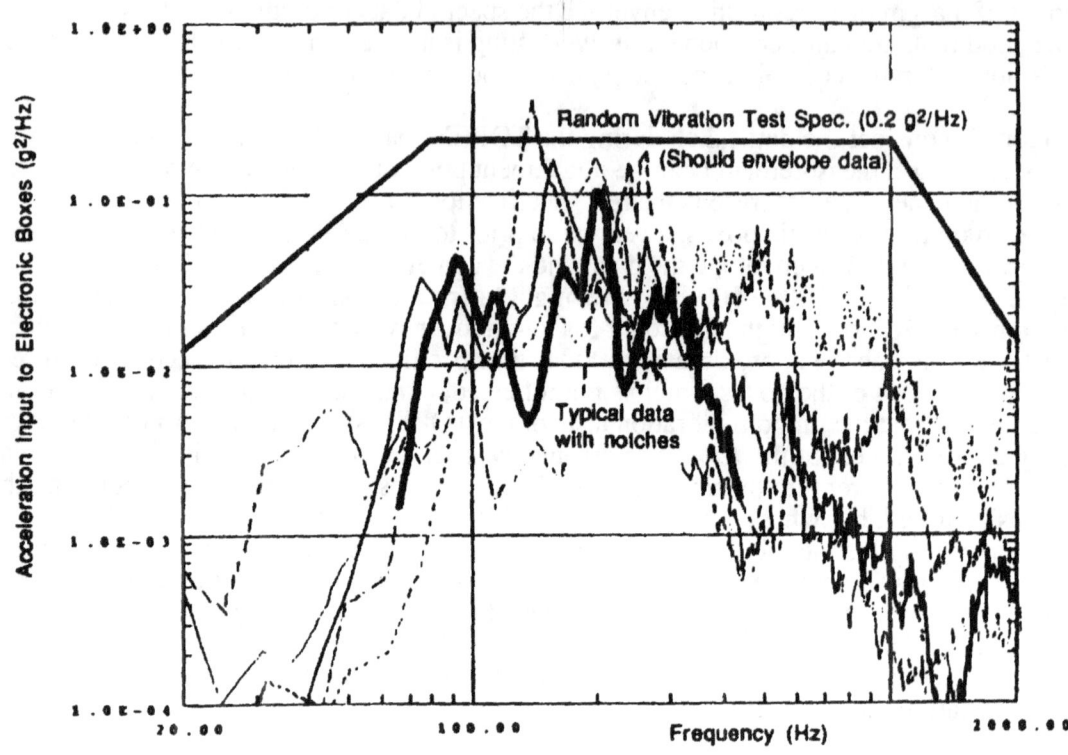

FIGURE 1. Measurements of Random Vibration Acceleration Spectra on Honeycomb Panel Near Mounting Feet of Electronic Boxes in TOPEX Spacecraft Acoustic Test

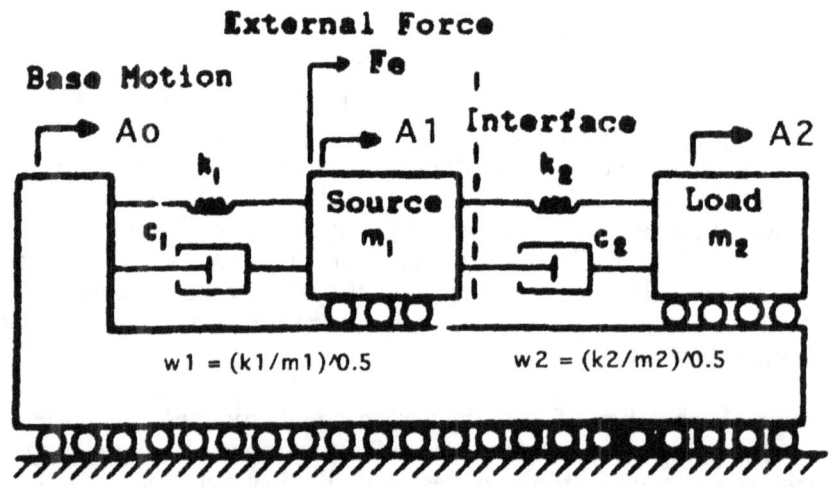

FIGURE 2. Simple Two-Degree-of-Freedom System (TDFS)directly excited oscillator. When the damping of the secondary oscillator is small rather than zero, the motion of the primary mass is small rather than zero.

force, Fig. 3b is the magnitude of the coupled system interface acceleration, and Fig. 3c is the magnitude of the load apparent mass.

In Fig. 3, notice first that the interface force and interface acceleration both have peaks at the two coupled system resonance frequencies of 0.62 f_o and 1.62 f_o. Notice further that the interface acceleration has a notch of depth Q^{-1} at the load fixed-base resonance frequency f_o, where and the load apparent mass has a peak of height Q. This notch in the interface acceleration is just the dynamic absorber effect. This example illustrates the general dynamic absorber result, that the frequency spectrum of acceleration at the interface between the spacecraft and instrument will have notches at the fixed-base resonance frequencies of the instrument.

The example in Fig. 3 may also be used to illustrate the overtesting resulting from enveloping the interface acceleration, and how force limiting will alleviate this overtesting. In the coupled system, the interface force peak of 80 at the lower resonance frequency of 0.62 f_o results from multiplying the interface acceleration peak of 50 by the load apparent mass value of 1.6. In a conventional vibration test without force limiting, the corresponding shaker force would be of 2500, which is the interface acceleration envelope of 50 times the load apparent mass peak of 50 at the load resonance frequency f_o. With force limiting, the input acceleration would be notched at the load resonance frequency f_o to reduce the shaker force by a factor of 2500/80 or 31.25.

3.2 Dual Control of Acceleration and Force

Conventional vibration tests are conducted by controlling only the acceleration input to the test item. In theory, if the frequency spectrum of the acceleration input in the test, including peaks and valleys, were identical to that of the interface acceleration in the flight mounting configuration, and if the boundary conditions for other degrees-of-freedom (rotations, etc.) were the same as in the flight configuration, then the interface forces and all the responses would be the same in the test as in flight. However, this is seldom the case, primarily because of the necessity of using a smoothed or an enveloped representation of the flight interface acceleration as the test input, and secondarily because of frequency shifts associated with the unrealistic restraint of other degrees-of-freedom by the shaker mounting. It has been found that the dual control of the acceleration and force input from the shaker alleviates the overtesting problem associated with conventional vibration tests using only acceleration control.

3.2.1 Thevinen and Norton's Equivalent Circuit Theorems

Consider a source, consisting of a voltage source in series with a source impedance, which is connected to a load [31]. If we adopt the mechanical analogy in which force is current and velocity is voltage (Unfortunately, this is the dual of the mechanical analogy used in reference[31].), Thevinen's equivalent circuit theorem may be stated in mechanical terms as:

$$A = A_o - F/\underline{M},\qquad(1)$$

where A is the source-load interface acceleration, A_o is the free acceleration (i.e., the acceleration that would exist at the interface if the load were removed), F is the interface force, and $\underline{M}$ is the source apparent mass measured at the interface. (Apparent mass is discussed in the next section.) All the terms in Equation 1 are complex and a function of frequency.

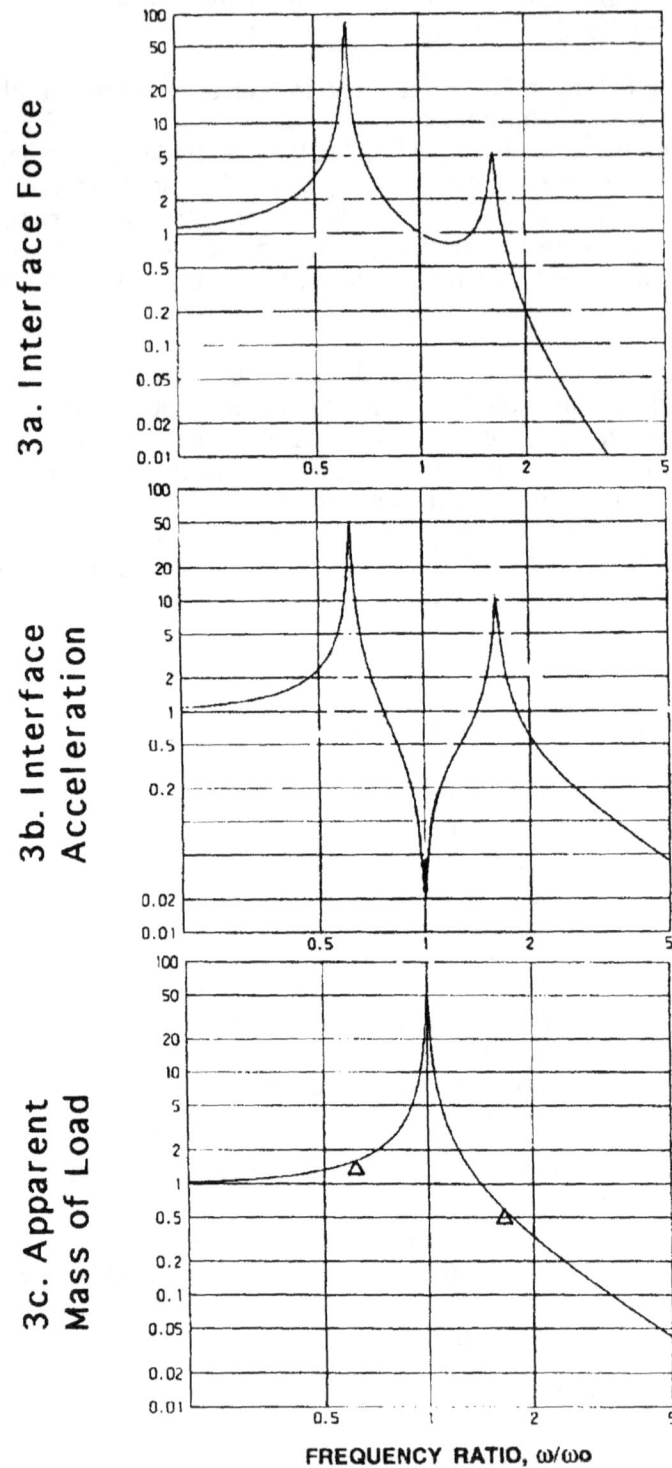

FIGURE 3. Interface Force, Interface Acceleration, and Load Apparent Mass FRF's for Simple TDFS in Figure 2 with $\omega_1 = \omega_2 = \omega_0$, $m_1 = m_2 = 1$, $A_0 = 1$ & $Q = 50$

Using the same electo-mechanical analogy, consider a source, consisting of a current source in parallel with a source impedance, connected to a load. Norton's equivalent circuit theorem stated in mechanical terms is :

$$F = F_o - A \underline{M},\qquad(2)$$

where F_o is the blocked force (i.e., the force that would be required at the interface to make the motion zero).

Equations (Eq.)1 and 2 may be manipulated to eliminate both F and A yielding the following relationship between the blocked force, free acceleration, and the source apparent mass:

$$F_o / A_o = \underline{M}.\qquad(3)$$

3.2.2 Dual Control Equations

Alternately, the source apparent mass $\underline{M}$ may be eliminated from Eqs. 1 and 2, to yield the following [20]:

$$1 = A/A_o + F/F_o,\qquad(4)$$

which provides a theoretical basis for dual control of vibration tests.

Equation 4 is exact but difficult to apply because the terms on the right hand side are complex and complicated functions of frequency. The phase of the inputs and the impedance are difficult to determine analytically or experimentally, although some exploratory work on this problem was conducted some 25 years ago [4,14]. Little recent work on exact mechanical impedance simulation is available, and most commercially available vibration test controllers can not control phase angle to a specification.

An alternative, approximate formulation for the control of vibration tests is provided by the following extremal Eqs. [7]:

$$|A|/|A_s| \le 1 \text{ and } |F|/|F_s| \le 1,\qquad(5)$$

in which A_s represents the acceleration specification and F_s represents the force specification. In Eq. 5, the free acceleration and blocked force of Eq. 4 are replaced by the corresponding specifications which envelope the interface acceleration and force in the coupled system. With extremal control, the shaker current is adjusted in each narrow frequency band so that the larger of the two ratios in Eq. 5 is equal to unity. At frequencies other than the test item resonances, the acceleration specification usually controls the test level; at the resonances, the base reaction force increases and the force specification limits the input.

Most vibration controllers have the capability for extremal control, but older controllers allow only one reference specification. To implement dual control in this case, a filter must be used to scale the shaker force feedback signal to an equivalent acceleration [17].New controllers allow separate specifications for limit channels, so Eq. 2 may be directly implemented. Force limiting has been used primarily for random vibration tests, but the application to swept sine tests is also practical and beneficial.

3.3 Structural Impedance Characterization

3.3.1 Apparent Mass

In this monograph structural impedance will be characterized as "apparent mass", which is the preferred name for the frequency response function (FRF) consisting of the ratio of reaction force to prescribed acceleration [32]. (Apparent mass symbols will be underscored in this monograph to distinguish them from other mass quantities.) The force and prescribed acceleration in the apparent mass usually refer to the same degree-of-freedom. (In the literature, this is often called the "drive point" as distinguished from the "transfer" apparent mass.) The accelerations at other boundary degrees-of-freedom should be constrained to be zero if one is dealing with a multiple drive point problem [8], but herein, only a single drive point is usually of interest, and this consideration is ignored. The apparent mass is generally a complex quantity, with magnitude and phase, but herein the term apparent mass will often be used in referring to only the magnitude. The apparent mass can vary greatly with frequency, as one passes through resonances. Therefore the apparent mass reflects the stiffness and damping characteristics of a structure, as well as the mass characteristics.

The closed form solution for the apparent mass of a rod excited at one end and free at the other end is given by [33]:

$$\frac{F(\omega)}{A(\omega)} = \underline{M}(\omega) = (i\rho_l c/\omega)\,(1+i\zeta)\,\frac{\tan(\pi\omega/2\omega_l) - i\,\tanh(\pi\zeta\omega/2\omega_l)}{1 + i\,\tan(\pi\omega/2\omega_l)\,\tanh(\pi\zeta\omega/2\omega_l)}, \qquad (6)$$

where: ρ_l is the mass per unit length, c is the speed of longitudinal waves $(EA/\rho_l)^{1/2}$ where E is Young's modulus and A is cross-section area, ζ is the critical damping ratio, and ω_l is the fundamental frequency $\pi c/2L$ with L the rod length. This result is plotted as the solid line in Fig. 4 for a critical damping ratio ζ of 2.5%.

3.3.2 Effective Mass

Another mass-like quantity of great significance in structural analysis and for impedance simulation is the "effective mass" [15,34]. A formal definition of the effective mass, which encompasses multiple degrees-of-freedom and off-diagonal terms, as well as equations which enable the effective mass to be calculated with NASTRAN, is given in the next section.

For the beam driven at one end, the apparent mass may also be expressed as a modal expansion involving the effective modal masses, m_n, and the single-degree-of-freedom frequency response factors:

$$\underline{M}(\omega) = M_o - \Sigma_n\, m_n\, \frac{(1 + i\,2\zeta)}{\{[(1 - (\omega/\omega_n)^2] + i\,2\zeta\}}, \quad n = 1,2,3\,\ldots\ldots \qquad (7)$$

3.3 Structural Impedance Characterization

Revised 9/27/00
Equation 7 only

3.3.1 Apparent Mass

In this monograph structural impedance will be characterized as "apparent mass", which is the preferred name for the frequency response function (FRF) consisting of the ratio of reaction force to prescribed acceleration [32]. (Apparent mass symbols will be underscored in this monograph to distinguish them from other mass quantities.) The force and prescribed acceleration in the apparent mass usually refer to the same degree-of-freedom. (In the literature, this is often called the "drive point" as distinguished from the "transfer" apparent mass.) The accelerations at other boundary degrees-of-freedom should be constrained to be zero if one is dealing with a multiple drive point problem [8], but herein, only a single drive point is usually of interest, and this consideration is ignored. The apparent mass is generally a complex quantity, with magnitude and phase, but herein the term apparent mass will often be used in referring to only the magnitude. The apparent mass can vary greatly with frequency, as one passes through resonances. Therefore the apparent mass reflects the stiffness and damping characteristics of a structure, as well as the mass characteristics.

The closed form solution for the apparent mass of a rod excited at one end and free at the other end is given by [33]:

$$\frac{F(\omega)}{A(\omega)} = \underline{M}(\omega) = (i\rho_1 c/\omega)\ (1+i\zeta)\ \frac{\tan(\pi\omega/2\omega_1) - i\tanh(\pi\zeta\omega/2\omega_1)}{1 + i\tan(\pi\omega/2\omega_1)\tanh(\pi\zeta\omega/2\omega_1)}, \quad (6)$$

where: ρ_1 is the mass per unit length, c is the speed of longitudinal waves $(EA/\rho_1)^{1/2}$ where E is Young's modulus and A is cross-section area, ζ is the critical damping ratio, and ω_1 is the fundamental frequency $\pi c/2L$ with L the rod length. This result is plotted as the solid line in Fig. 4 for a critical damping ratio ζ of 2.5%.

3.3.2 Effective Mass

Another mass-like quantity of great significance in structural analysis and for impedance simulation is the "effective mass" [15,34]. A formal definition of the effective mass, which encompasses multiple degrees-of-freedom and off-diagonal terms, as well as equations which enable the effective mass to be calculated with NASTRAN, is given in the next section.

For the beam driven at one end, the apparent mass may also be expressed as a modal expansion involving the effective modal masses, m_n, and the single-degree-of-freedom frequency response factors:

$$\underline{M}(\omega) = \Sigma_n\ m_n\ \frac{(1 + i2\zeta)}{\{[(1 - (\omega/\omega_n)^2] + i2\zeta\}}, \quad n = 1,2,3 \ldots\ldots \quad (7)$$

3-8

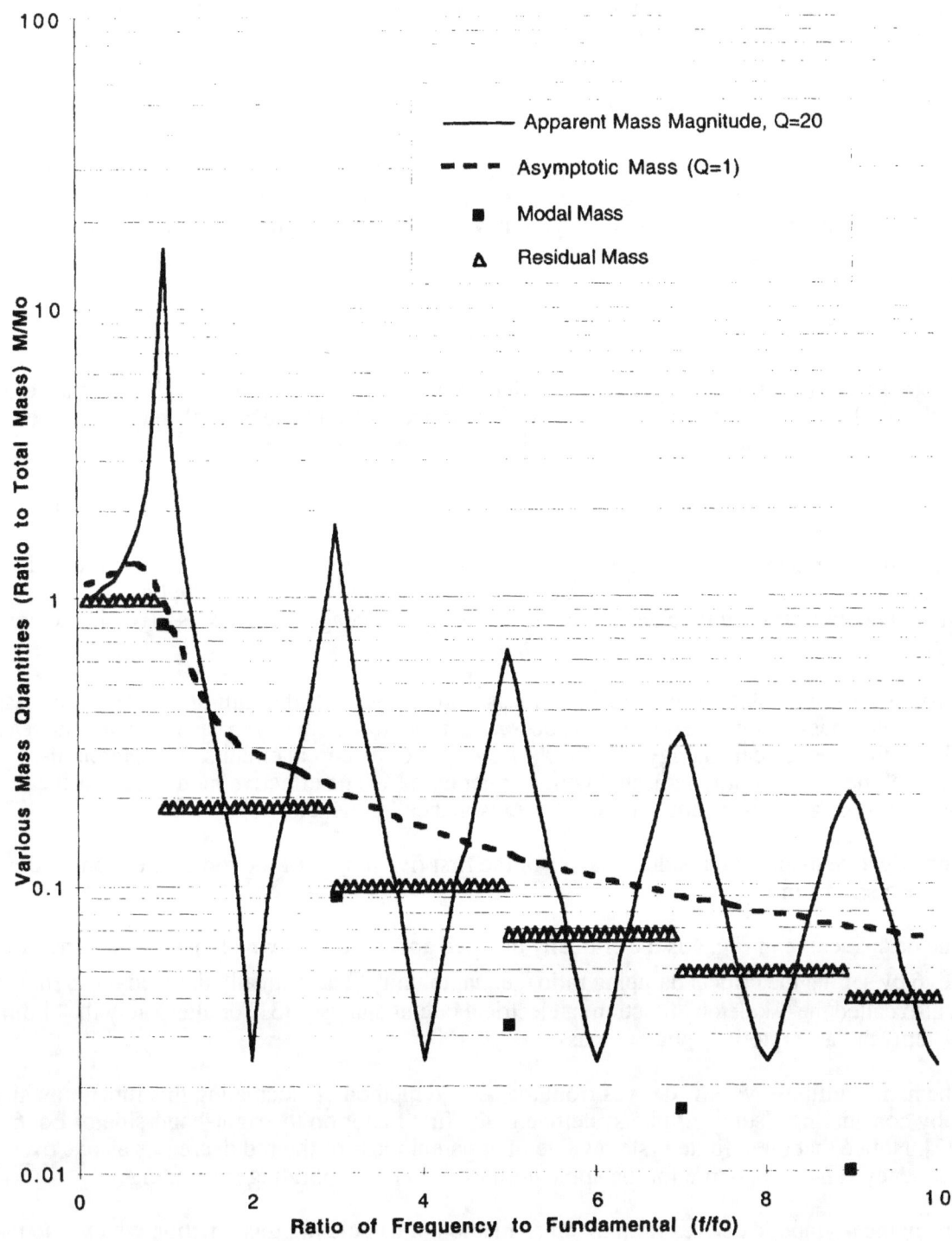

FIGURE 4. Apparent Mass, Asymptotic Mass, Modal Mass, and Residual Mass of
Longitudinally Vibrating Rod, Excited at One End and Free at Other End

where the structural form of damping has been assumed, M_o is the total mass $\rho_l L$, and m_n is the effective mass of the nth mode which is given by [34]:

$$m_n = 8 M_o / [\pi^2 (2n - 1)^2], \tag{8}$$

where ω_n is the natural frequency of the nth mode which is equal to $\omega_l (2n - 1)$. Equation 7 may be viewed as a definition of the drive point effective mass. The sum of the effective masses over all the modes is the total mass, which may be verified by summing the rod effective masses, given by Eq. 8, over all n [34].

3.3.3 Residual Mass

Another mass quantity closely related to effective mass is the "residual mass", which is defined as the total mass minus the effective mass of the modes which have natural frequencies below the excitation frequency. Thus the residual mass of the Nth mode is:

$$M (N) = M_o - \Sigma_{n = 1 \text{ to } N} m_n. \tag{9}$$

The residual mass may be interpreted as the fraction of the total mass which moves with the input acceleration, like a rigid body. A more precise definition of the residual mass concept, which encompasses multiple degrees-of-freedom and off-diagonal terms, is given in the next section.

It follows from the definition of residual mass, and the fact that the sum of the effective masses is the total mass, that the residual mass decreases monotonically to zero as frequency increases. This is the mechanical analogy of Foster's Theorem for electrical circuits [31]. Herein the residual mass is generally indicated with a upper case M. The effective modal mass is the negative change in the residual mass, at the resonance frequencies.

The effective modal and residual masses of the first five modes of the rod excited longitudinally are also shown in Fig. 4.

The dashed curve in Fig. 4 is the critically damped apparent mass which may be obtained from Eq. 6 by setting the critical damping ratio ζ equal to unity. The critically damped apparent mass is also called the "skeleton" function in electrical circuit analyses [31] or alternately the "infinite system" or "asymptotic" apparent mass.

The name "infinite system" derives from the second method of calculating this function which is by considering a semi-infinite system, e.g. the first factor on the right-hand-side of Eq. 6 [33]. Notice that the infinite system value of apparent mass of the rod decreases as one over frequency. This is also true for the apparent mass of a plate vibrating in bending.

The name asymptotic derives from the third method of calculating this function which is to take a geometric average of the apparent mass FRF over frequency, so that there is equal area above and below the curve on a log-log plot [35, 36]. (Observe the equal area characteristic of the critically damped apparent mass in Fig. 4.) The asymptotic form of the apparent mass is very important to the development herein, because it will be used to represent experimental apparent mass data measured either in tap tests or shaker tests. Notice in Fig. 4 that the asymptotic apparent mass is a generous envelope of the residual mass. The asymptotic apparent mass, which must include stiffness as well as mass contributions, is approximately equal to $2^{1/2}$ times the residual mass at the natural frequencies in Fig. 4. Herein the asymptotic apparent mass will be used as an approximation to the residual mass.

3.3.4 FEM Calculation of Effective Mass

Subdividing the displacement vector into unrestrained absolute displacements u_f and prescribed absolute displacements, u_p the equilibrium Equation is [15, 34]:

$$
\begin{bmatrix} m_{FF} \mid m_{FP} \\ \overline{} \\ m_{PF} \mid m_{PP} \end{bmatrix} \left\{ \begin{matrix} d^2u_F/dt^2 \\ \overline{} \\ d^2u_P/dt^2 \end{matrix} \right\} + \begin{bmatrix} k_{FF} \mid k_{FP} \\ \overline{} \\ k_{PF} \mid k_{PP} \end{bmatrix} \left\{ \begin{matrix} u_F \\ -- \\ u_P \end{matrix} \right\} = \left\{ \begin{matrix} f_F \\ -- \\ f_P \end{matrix} \right\}.
\tag{10}
$$

Let: $\{u\} = \phi\, U = \dfrac{[\,\phi_N \mid \phi_P\,]}{[\,0 \mid I_{PP}\,]} \left\{ \begin{matrix} U_N \\ --- \\ U_P \end{matrix} \right\},$

For correct printing of Eqs.10-12
see Eqs. B1-B3 in Appendix C (11)

where ϕ_N are normal modes and ϕ_R are rigid body modes associated with a kinematic set of unit prescribed motions, and U_N is the generalized modal relative displacement and U_P is the generalized prescribed absolute displacement. Substituting and pre-multiplying by ϕ^T yields:

$$
\begin{bmatrix} M_{NN} \mid M_{NP} \\ \overline{} \\ M^T_{NP} \mid M_{PP} \end{bmatrix} \left\{ \begin{matrix} d^2U_N/dt^2 \\ \overline{} \\ d^2U_P/dt^2 \end{matrix} \right\} + \begin{bmatrix} \omega^2_N M_{NN} \mid 0 \\ \overline{} \\ 0 \mid 0 \end{bmatrix} \left\{ \begin{matrix} U_N \\ -- \\ U_P \end{matrix} \right\} \begin{matrix} F_F \\ \\ F_P \end{matrix} = \{----\},
\tag{12}
$$

where: $\quad M_{NN} = \phi_N{}^T m_{FF}\, \phi_N$ (13)

$M_{NP} = \phi_N{}^T m_{FF}\, \phi_P + \phi_N{}^T m_{FP}\, I_{PP}$ (14)

$M_{PP} = I_{PP}\, m_{PP}\, I_{PP} + I_{PP}\, m_{PF}\, \phi_P + \phi_P{}^T m_{FP}\, I_{PP} + \phi_P{}^T m_{FF}\, \phi_P$ (15)

$F_P = I_{PP}\, f_P$. (16)

For: $\quad d^2U_P/dt^2 = U_P = F_F = 0,\ d^2U_n/dt^2 = -\omega^2 U_n,$ and for $U_n = 1$:

$$
M_{nP}{}^T = -\, F_P / \omega_n{}^2,
\tag{17}
$$

where n indicates a single mode. (Note that $M_{nP}{}^T$ is in mass units.) M_{nP} is sometimes called the elastic-rigid coupling or the modal participation factor for the nth mode. If the model is restrained at a single point, the reaction (F_P) in Eq. 17 is the SPCFORCE at that point in a NASTRAN modal analysis.

The initial value of M_{PP} is the rigid body mass matrix. If a Gaussian decomposition of the total modal mass in Eq.12 is performed, it subtracts the contribution of each normal mode, called the effective mass:

$$
M_{nP}{}^T M_{nn}{}^{-1} M_{nP} ,
\tag{18}
$$

from the current $M_{PP}{}^n$, which is the residual mass after excluding the mass associated with the already processed n modes.

Consider the ratio of the reaction force in a particular direction p, to the prescribed acceleration in a particular direction q; the effective mass, $M_{np}^T M_{nn}^{-1} M_{nq}$, is the same as the contribution of the nth mode to this ratio, divided by the single-degree-of-freedom frequency response factor. Please note that the values of the effective mass are independent of the modal normalization.

Generally the reaction is desired in the same direction as the excitation and the effective mass for the common direction is a diagonal of the $M_{nP}^T M_{nn}^{-1} M_{nP}$ 6x6 matrix, and the residual mass for that element monatonically decreases as more and more modes are processed. The sum of the common-direction effective masses for all modes is equal to the total mass, or moment of inertia for that direction. If there is no common direction, the foregoing is not true. If $m_{FP} = m_{PP}=0$ the residual mass after processing a complete set of modes is a 6x6 null matrix. If m_{FP} and m_{PP} are not equal to zero, the value of M_{PP}^N after processing a complete set of modes is: $m_{PP} - m_{FP}^T \phi_N M_{NN}^{-1} \phi_N^T m_{FP}$, which must be positive definite.

The highest reaction of a single mode for a given excitation level may not occur along one of the axes used in analysis or test. The highest reaction force (not moment) will occur for excitation along an axis such that its direction cosines are proportional to the diagonal terms of the effective mass along the analysis axis. The effective mass along this axis is $(M_{1n}^2 + M_{2n}^2 + M_{3n}^2) / M_{nn}$.

4.0 Force Limits

There are virtually no flight data and little system test data on the vibratory forces at mounting structure and test item interfaces. Currently force limits for vibration tests are therefore derived using one of three methods: 1) calculated using two-degree-of freedom or other analytical models of the coupled source/load system together with measured or FEM effective mass data for the mounting structure and test item, 2) estimated using a semi-empirical method based on system test data and heuristic arguments, or 3) taken from the quasi-static design criteria which may be based on coupled loads analysis or a simple mass acceleration curve. In the first two methods, which are usually applicable in the random vibration frequency regime, the force specification is based on and proportional to the conventional acceleration specification. Any conservatism (or error) in the acceleration specification carries over to the force specification. In the third method, which is usually limited to static and low frequency sine-sweep or transient vibration tests, the force limit is derived independently from the acceleration specification.

4.1 Coupled System Methods

The basic approach to calculating force limits with a coupled system model involves four steps:

1. Development of a parametric model of the coupled source and load system which might be an FEM or a multiple degree-of-freedom modal model,

2. Identification of the model parameters using measured apparent mass or FEM modal frequency and effective mass information,

3. Solution of the coupled system problem to obtain the ratio of the force frequency envelope to the acceleration envelope at the source/load interface, and

4. Multiplication of the ratio of envelopes by the acceleration specification to obtain an analogous force specification.

In the following two sections, this basic method of calculating force limits using a coupled source/load model is implemented for two specific cases where the coupled model is a simple and a complex two-degree-of-freedom system (TDFS).

For both the flight configuration with a coupled source and load and the vibration test configuration with an isolated load, the interface force spectral density S_{FF} is related to the interface acceleration spectral density S_{AA} as [27]:

$$S_{FF}(\omega) = |\underline{M}_2(\omega)|^2 \, S_{AA}(\omega). \tag{19}$$

The load apparent mass $\underline{M}_2$ is a frequency response function (FRF) which includes mass, damping, and stiffness effects. The frequency dependence is shown explicitly in Eq. 19 to emphasize that the relation between force and acceleration applies at each frequency.

It can be shown [37] that, for white noise base motion or external force excitation of the coupled system in Fig. 2, the interface acceleration and force spectral densities both peak at the same frequencies, i.e. the coupled system natural frequencies. The load apparent mass, evaluated at one of these natural frequencies, may be interpreted as the ratio of the force spectral peak to the acceleration spectral peak at that natural frequency.

Both the simple and the complex TDFS methods are derived by multiplying the conventional acceleration specification, which is assumed to properly envelope the acceleration spectral peaks, by the load apparent mass, evaluated at the coupled system resonance frequencies. A central point of this approach is that the load apparent mass must be evaluated at the coupled system, or shifted, resonance frequencies. The values of the load apparent mass at the coupled system resonance frequencies are considerably less than the peak value at the load uncoupled resonance frequency.

Fig. 3 illustrates the application of Eq. 19 to the simple, TDFS model shown in Fig. 2 when the oscillators are identical. Fig. 3a is the magnitude of the coupled system interface force, which is equal to the product of the load apparent mass in Fig. 3c and the interface acceleration in Fig. 3b.

4.1.1 Simple TDFS Method

The force limit is here calculated for the simple, non-identical TDFS in Fig. 2 with different masses of the source and the load oscillators [27]. For this TDFS, the maximum response of the load and therefore the maximum interface force occur when the uncoupled resonance frequency of the load equals that of the source [38]. For this case, the characteristic equation is that of a classical dynamic absorber[30]:

$$(\omega/\omega_o)^2 = 1+(m_2/m_1)/2 \pm [(m_2/m_1)+(m_2/m_1)^2/4)]^{0.5} . \tag{20}$$

The ratio of the interface force to acceleration spectral densities, calculated as in Eq. 19 from the magnitude squared of the load apparent mass, is:

$$S_{FF}/(S_{AA}\ m_2^2) = [1+(\omega/\omega_o)^2/Q_2^2] /\{[1-(\omega/\omega_o)^2]^2 +(\omega/\omega_o)^2/Q_2^2\}. \tag{21}$$

The force spectral density, normalized by the load mass squared and by the acceleration spectral density, at the two coupled system resonances is obtained by combining Eqs. 20 and 21. For this TDFS the normalized force is just slightly larger at the lower resonance frequency of Eq. 20. The maximum normalized force spectral density, obtained by evaluating Eq. 21 at the lower resonance frequency from Eq. 20, is plotted against the ratio of load to source mass for three values of Q_2 in Fig. 5.

In Fig. 5, for very small (0.0001) values of the ratio of load to source mass, the load has little effect on the source, and the maximum normalized force approaches Q squared. For larger ratios of the masses, the maximum force is smaller because of the vibration absorber effect at the load resonance frequency. For equal load and source masses, the maximum normalized force in Fig. 5 is 2.56 or $(1.6)^2$, as in the numerical example of Fig. 3.

Use of Fig. 5 to define force specifications requires that the oscillator masses in Fig. 2 be deduced from the properties of the distributed source and load systems. Clearly the oscillator masses must vary with frequency, and it has proven convenient to define them, as well as the resulting force specifications, in one-third octave bands. One might think that the oscillator masses should be identified with the resonant modal or effective masses of the distributed system. However, this choice results in no force at frequencies where the system lacks resonances. A more conservative approach is to identify the oscillator masses in Fig. 5 with the residual masses of the distributed systems.

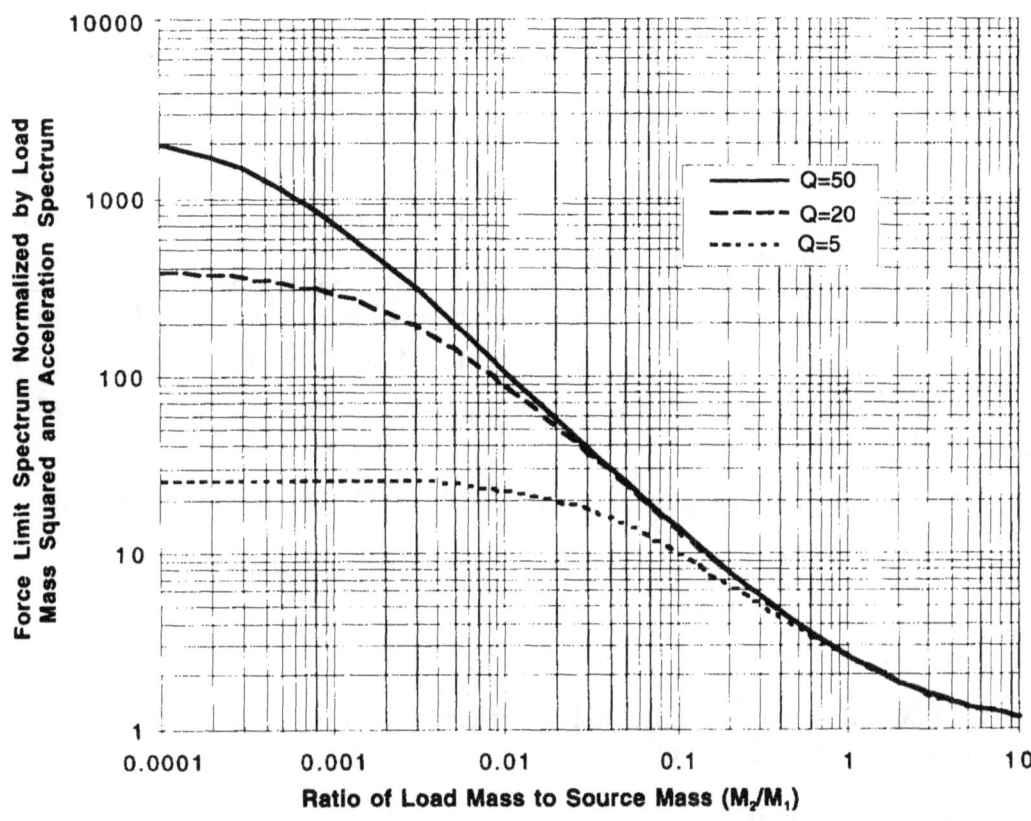

FIGURE 5. Random Vibration Force Specification Calculated From Simple TDFS

Defining the oscillator masses as residual masses instead of as modal masses, over estimates both the load and source masses which over estimates the interface forces calculated using Fig. 5. This approach is conservative for testing, but it is not very conceptually satisfying. The Fig. 2 model, with only one mass for the source and another for the load, is basically deficient in that it can not represent the force contributions of both resonant and non-resonant structural modes. This leads to the consideration of a more complex TDFS model, which for some configurations predicts even higher force limits than the simple TDFS model.

4.1.2 Complex TDFS Method

Here the force limit is calculated for a more complex TDFS model in which the source and load each have two masses to represent both the residual and modal masses of a continuous system [27]. As in the case of the simple TDFS, it is assumed that the acceleration specification correctly envelopes the higher of the two acceleration peaks of the coupled source and load system. However, for the more complex TDFS, the calculation of the force limit requires evaluating the relative sizes of the acceleration peaks at the two coupled system resonance frequencies (i.e. determining the mode shapes of the coupled system), which requires some specific assumptions about how the system is excited. Calculation of the force limit for this system also necessitates a tuning analysis, in which the maximum force is calculated for different ratios of the load and source uncoupled resonance frequencies. The complexity of the model requires that the results be presented in tabular form for different ratios of modal to residual mass, for both the source and the load.

Fig. 6a shows a model of a source or a load in which each mode may be represented as a collection of single-degree-of-freedom systems attached in parallel to a common interface. (This type of model is sometimes called an "asparagus patch" model [34].) The masses of the single-degree-of freedom systems in this parallel oscillator model are defined by the effective masses of the modes of the distributed system.

When this parallel oscillator model is excited at the interface with a frequency near the resonance frequency ω_n of the nth mode, the model may be simplified to that in Fig. 6b, where m_n is the modal mass of the nth mode and M_n is the residual mass, i.e. the sum of the masses of all modes with resonance frequencies above the excitation frequency. Finally, Fig. 6c shows a coupled system with a residual and modal mass model of both the source and the load. The ratio of modal to residual mass is $\alpha_1=m_1/M_1$ for the source and $\alpha_2=m_2/M_2$ for the load; the ratio of load and source uncoupled resonance frequencies is $\Omega=\omega_2/\omega_1$; and the ratio of load and source residual masses is $\mu=M_2/M_1$.

The undamped resonance frequencies of the coupled system in Fig. 6c are solutions of:

$$(1-\beta_1^2)(1-\beta_2^2)+ \alpha_1(1-\beta_2^2) + \mu(1-\beta_1^2)(1-\beta_2^2)+\mu\alpha_2(1-\beta_1^2) = 0, \qquad (22)$$

with $\beta_1=\omega/\omega_1$, $\beta_2=\omega/\omega_2$, $\omega_1=(k_1/m_1)^{0.5}$, and $\omega_2=(k_2/m_2)^{0.5}$.

Using Ω to eliminate β_1, the two undamped resonance frequencies of the coupled system are found from the quadratic equation solution:

$$\beta_2^2 = -B/2 \pm (B^2-4C)^{0.5}/2, \qquad (23)$$

where:

$$B = -[(1+\mu+\alpha_1)/\Omega^2+(1+\mu+\mu\alpha_2)]/(1+\mu) \text{ and } C = (1+\mu+\alpha_1+\mu\alpha_2)/[(1+\mu)\Omega^2].$$

The interface force spectral density, normalized by the acceleration spectral density and the load residual mass squared, calculated as in Eq. 19 from the magnitude squared of the load dynamic mass, is:

$$|\underline{M}_2|^2/M_2^2 = \{[(1-\beta_2^2)+\alpha_2]^2+\beta_2^2(1+\alpha_2)^2/Q_2^2\}/ [(1-\beta_2^2)^2+\beta_2^2/Q_2^2] . \qquad (24)$$

Combining Eqs. 23 and 24, yields the normalized force spectral density at each of the two coupled system resonance frequencies.

The desired result is the ratio of the larger of the two force spectral density peaks to the larger of the two acceleration spectral density peaks, the former being the desired force limit and the latter corresponding to the acceleration specification. Unfortunately, the peak acceleration and peak force do not necessarily occur at the same frequency, e.g. the peak acceleration may be at the higher of the two coupled system resonance frequencies while the peak force is at the lower of the two frequencies. This, in fact, occurs when the uncoupled resonance frequencies of the load and source are approximately equal, that is for Ω near unity.

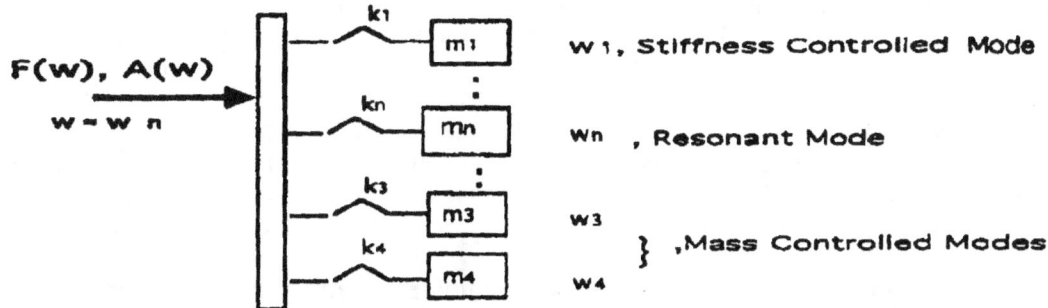

FIG. 6A. ASPARAGUS PATCH MODEL OF SOURCE OR LOAD

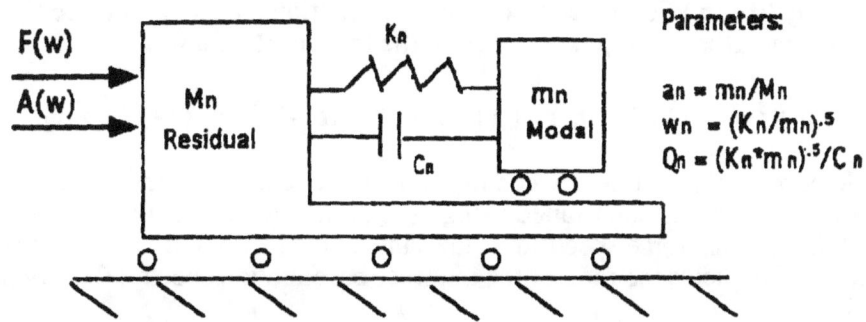

FIG. 6B. RESIDUAL AND MODAL MASS MODEL OF SOURCE OR LOAD

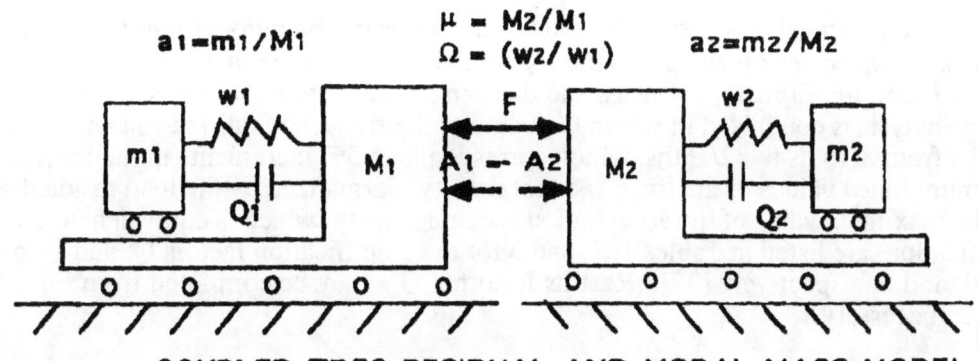

FIG. 6C. COUPLED TDFS RESIDUAL AND MODAL MASS MODEL

FIGURE 6. Complex Two-Degree-of-Freedom-System (TDFS)

To obtain the desired result, it is necessary to calculate the ratio of the two acceleration spectral density peaks of the coupled system, and this ratio depends on how the system is excited. Herein, it is assumed that the modal mass of the source is excited by an external force with a flat spectral density over the frequency band including the two resonance frequencies of the coupled system. (Two other excitation possibilities would be that the spectrum of the free acceleration or the blocked force of the residual mass of the source is constant with frequency.) The flat external force acting on the source modal mass is chosen, because it is thought to be the most typical and because it yields the highest force limits when the load and source masses are comparable.

For any excitation of the source system, the magnitude squared of the ratio of interface acceleration A to the free acceleration A_{lo} of the residual mass of the source is:

$$|A/A_{lo}|^2 = |\underline{M}_1/(\underline{M}_1+\underline{M}_2)|^2. \tag{25}$$

(Note that the evaluation of Eq. 25 requires the complete complex forms of $\underline{M}_1$ and $\underline{M}_2$ which are not given explicitly herein.) For the chosen form of excitation, an external force F_e acting on the source modal mass m_1, the magnitude squared of the free acceleration A_{lo} is:

$$|A_{lo}/(F_e/m_1)|^2 = \beta_1^4(1+\beta_1^2/Q_1^2) / \{[(1-\beta_1^2)(1-\beta_1^2/\alpha_1)-1]^2 + \beta_1^6(1+1/\alpha_1)^2/Q_1^2\}. \tag{26}$$

The free acceleration is eliminated between Eqs. 25 and 26, and the interface acceleration at the two coupled system resonances is determined using Ω to eliminate β_1 and substituting β_2^2 from Eq. 23. Assuming that the external force spectrum is the same at the lower and upper resonance frequencies of the coupled system, yields the ratio of the interface acceleration spectral density peaks at these two frequencies.

The dynamic mass in Eq. 24 is scaled by multiplying the dynamic mass at the resonance frequency corresponding to the smaller acceleration peak by the ratio of the smaller to the larger acceleration peak and by multiplying the dynamic mass at the other frequency by unity. Finally, the larger of the two thus scaled dynamic masses is used as the ratio of the greater force spectral density peak to the greater acceleration spectral density peak.

The final step in the derivation of the force limit is to vary the ratio, $\Omega = \omega_2/\omega_1$, of the uncoupled resonance frequencies of the load to the source to insure that the maximum value of the interface force is found for all mass, stiffness, and damping combinations for the system in Fig. 6c. A tuning analysis is conducted in which the value of the frequency ratio Ω squared is varied by 1/16ths from 8/16ths to 32/16ths, which corresponds to 3% increments in the frequency ratio. The maximum tuned values of the force spectral density, normalized by the load residual mass squared and the maximum value of the acceleration spectral density (which is equivalent to the acceleration specification) are listed in Tables 1, 2, and 3 for the amplification factors Q_1 and Q_2 both equal to 50, 20, and 5, respectively [37]. (Results for other Q's may be computed from Eqs. 22 -26, if deemed necessary.)

The normalized force spectral density in Tables 1-3 is unity when $\alpha_2 = 0$ (see Eq. 24) and should be interpolated for $0.125 < \alpha_2 < 0$. It is suggested that the force spectral density for $\alpha_1 < 0.25$, be taken as the value at $\alpha_1 = 0.25$. (Small α_1's correspond to local source modes, which may not be relevant to the interface environment. Also, the interface force for no source modal mass is different than the asymptotic value for $\alpha_1 = 0$, which corresponds to a very small mass moving at very large amplitude.)

Table 1. Force Limit Spectrum for Complex TDFS with Q=50
(Normalized By Load Residual Mass Squared and Acceleration Spectrum)

Ratio of modal to residual mass m1/M1, m2/M2	Residual mass ratio , M2/M1								
	0.001	0.003	0.01	0.03	0.1	0.3	1	3	10
8.0, 8.0	949	950	954	966	1021	1206	1268	1261	1265
8.0, 4.0	237	237	238	239	244	260	299	270	253
8.0, 2.0	59	59	59	59	60	61	69	73	69
8.0, 1.0	15	15	15	15	15	15	4	7	6
8.0, 0.5	4	4	4	4	4	4	4	2	5
8.0, 0.25	1	1	1	1	1	1	1	1	3
8.0, 0.125	1	1	1	1	1	1	1	1	1
8.0, 0.0	1	1	1	1	1	1	1	1	1
4.0, 8.0	884	880	870	860	916	1058	1086	1134	1255
4.0, 4.0	221	221	220	219	223	253	257	253	255
4.0, 2.0	55	55	55	56	57	62	73	69	67
4.0, 1.0	14	14	14	14	15	16	22	23	22
4.0, 0.5	3	3	3	4	4	4	6	10	10
4.0, 0.25	1	1	1	1	1	1	2	5	5
4.0, 0.125	1	1	1	1	1	1	1	3	3
4.0, 0.0	1	1	1	1	1	1	1	1	1
2.0, 8.0	1640	1521	1286	1075	1003	965	996	1119	1234
2.0, 4.0	420	404	364	311	275	261	240	241	257
2.0, 2.0	106	105	99	90	80	82	70	66	63
2.0, 1.0	27	27	26	25	24	26	25	23	22
2.0, 0.5	7	7	7	7	7	9	11	10	10
2.0, 0.25	2	2	2	2	1	3	6	5	6
2.0, 0.125	1	1	1	1	1	1	3	3	4
2.0, 0.0	1	1	1	1	1	1	1	1	1
1.0, 8.0	16554	6508	2921	1510	976	909	998	1114	1212
1.0, 4.0	7333	2965	1200	583	336	249	235	240	252
1.0, 2.0	3080	1345	502	248	128	84	71	67	64
1.0, 1.0	1189	592	229	112	53	34	26	23	24
1.0, 0.5	415	245	106	51	26	16	12	11	11
1.0, 0.25	132	94	48	23	13	9	6	6	6
1.0, 0.125	39	33	21	11	7	5	4	4	4
1.0, 0.0	1	1	1	1	1	1	1	1	1
0.5, 8.0	24199	9798	3726	1761	1046	887	994	1112	1202
0.5, 4.0	10238	4417	1672	738	368	249	229	242	248
0.5, 2.0	4046	1927	747	319	143	89	72	65	65
0.5, 1.0	1454	804	335	142	62	40	27	25	23
0.5, 0.5	472	311	148	66	30	18	13	12	10
0.5, 0.25	141	110	63	31	15	9	8	7	7
0.5, 0.125	40	36	26	15	8	5	4	5	5
0.5, 0.0	1	1	1	1	1	1	1	1	1
0.25, 8.0	33910	13269	4455	1996	1026	839	955	1111	1196
0.25, 4.0	14189	6185	2155	885	393	251	227	244	247
0.25, 2.0	5342	2736	1043	405	182	96	71	66	66
0.25, 1.0	1764	1111	492	205	80	45	28	23	22
0.25, 0.5	529	396	219	104	45	23	15	12	11
0.25, 0.25	149	128	85	47	22	12	8	8	7
0.25, 0.125	41	39	31	20	11	6	5	5	5
0.25, 0.0	1	1	1	1	1	1	1	1	1
0.125, 8.0	48146	18637	6361	2411	1072	855	936	1111	1194
0.125, 4.0	19122	8823	2885	1174	411	268	230	244	246
0.125, 2.0	6642	3788	1454	508	193	106	74	67	66
0.125, 1.0	2045	1434	684	271	105	48	30	24	22
0.125, 0.5	574	477	291	139	52	27	14	13	11
0.125, 0.25	155	142	110	66	31	15	9	7	7
0.125, 0.125	41	41	36	27	16	10	6	5	5
0.125, 0.0	1	1	1	1	1	1	1	1	1
0.0, 8.0	134767	66196	13561	2836	1136	874	917	1110	1191
0.0, 4.0	37885	28769	8669	1827	464	258	233	245	245
0.0, 2.0	9820	8998	5015	1203	276	110	69	68	66
0.0, 1.0	2484	2419	1962	823	187	54	30	25	22
0.0, 0.5	625	619	580	402	111	35	17	12	11
0.0, 0.25	157	157	154	136	69	25	10	8	7
0.0, 0.125	40	40	40	38	30	14	9	5	4
0.0, 0.0	1	1	1	1	1	1	1	1	1

Table 2. Force Limit Spectrum for Complex TDFS with Q=20
(Normalized By Load Residual Mass Squared and Acceleration Spectrum)

Ratio of modal to residual mass m1/M1, m2/M2	Residual mass ratio, M2/M1								
	0.001	0.003	0.01	0.03	0.1	0.3	1	3	10
8.0, 8.0	932	933	936	948	1001	1180	1240	1234	1238
8.0, 4.0	233	233	233	235	239	256	294	265	250
8.0, 2.0	58	58	58	58	59	60	68	73	68
8.0, 1.0	15	15	15	15	15	15	17	23	22
8.0, 0.5	4	4	4	4	4	4	4	7	6
8.0, 0.25	1	1	1	1	1	1	1	2	5
8.0, 0.125	1	1	1	1	1	1	1	1	3
8.0, 0.0	1	1	1	1	1	1	1	1	1
4.0, 8.0	871	867	858	849	904	1042	1067	1110	1229
4.0, 4.0	218	218	217	216	220	250	254	250	252
4.0, 2.0	55	55	55	55	56	61	72	68	67
4.0, 1.0	14	14	14	14	14	16	21	23	22
4.0, 0.5	3	3	4	4	4	4	6	10	10
4.0, 0.25	1	1	1	1	1	1	2	5	5
4.0, 0.125	1	1	1	1	1	1	1	3	3
4.0, 0.0	1	1	1	1	1	1	1	1	1
2.0, 8.0	1586	1478	1260	1061	990	946	982	1099	1201
2.0, 4.0	406	391	355	305	272	259	238	236	254
2.0, 2.0	103	101	97	88	79	82	70	65	62
2.0, 1.0	26	26	26	25	24	25	25	23	22
2.0, 0.5	7	7	7	7	7	9	10	10	10
2.0, 0.25	2	2	2	2	2	3	5	5	6
2.0, 0.125	1	1	1	1	1	1	3	3	4
2.0, 0.0	1	1	1	1	1	1	1	1	1
1.0, 8.0	11041	5731	2714	1486	967	901	984	1095	1181
1.0, 4.0	3869	2206	1105	567	332	247	233	238	248
1.0, 2.0	1228	826	432	226	125	83	71	66	64
1.0, 1.0	359	283	166	100	50	34	26	23	23
1.0, 0.5	100	89	63	42	24	15	12	11	11
1.0, 0.25	28	27	23	17	11	8	6	6	6
1.0, 0.125	8	8	8	7	5	5	4	4	4
1.0, 0.0	1	1	1	1	1	1	1	1	1
0.5, 8.0	13889	7720	3501	1726	1023	880	974	1093	1171
0.5, 4.0	4516	2895	1417	695	357	247	225	240	244
0.5, 2.0	1346	1003	561	283	136	89	70	64	65
0.5, 1.0	377	319	211	117	59	39	27	24	22
0.5, 0.5	102	95	74	48	27	17	12	11	10
0.5, 0.25	28	27	25	19	13	8	7	6	6
0.5, 0.125	8	8	8	8	6	5	4	4	4
0.5, 0.0	1	1	1	1	1	1	1	1	1
0.25, 8.0	17378	9978	4092	1944	1017	833	936	1092	1166
0.25, 4.0	5194	3725	1805	812	380	249	225	241	242
0.25, 2.0	1455	1205	741	359	173	93	71	66	65
0.25, 1.0	391	354	269	160	74	43	28	23	22
0.25, 0.5	103	99	86	63	38	22	14	12	11
0.25, 0.25	28	28	27	23	16	10	8	7	7
0.25, 0.125	8	8	8	8	7	5	5	4	4
0.25, 0.0	1	1	1	1	1	1	1	1	1
0.125, 8.0	19966	12425	5389	2331	1048	849	918	1092	1163
0.125, 4.0	5748	4417	2241	1080	400	266	228	242	241
0.125, 2.0	1533	1368	901	429	184	102	72	66	65
0.125, 1.0	400	380	312	192	91	45	29	24	22
0.125, 0.5	104	102	95	75	42	24	14	12	11
0.125, 0.25	27	28	27	26	20	13	8	7	6
0.125, 0.125	8	8	8	8	8	7	5	4	4
0.125, 0.0	1	1	1	1	1	1	1	1	1
0.0, 8.0	25114	21284	10111	2700	1125	867	900	1091	1161
0.0, 4.0	6394	6108	4156	1560	454	256	231	240	240
0.0, 2.0	1608	1590	1409	757	257	109	68	67	66
0.0, 1.0	404	403	390	310	148	52	30	25	22
0.0, 0.5	102	102	101	95	60	30	16	12	11
0.0, 0.25	27	27	27	26	22	17	9	7	6
0.0, 0.125	8	8	8	7	7	6	6	5	4
0.0, 0.0	1	1	1	1	1	1	1	1	1

Table 3. Force Limit Spectrum for Complex TDFS with Q=5
(Normalized By Load Residual Mass Squared and Acceleration Spectrum)

Ratio of modal to residual mass m1/M1, m2/M2	Residual mass ratio , M2/M1								
	0.001	0.003	0.01	0.03	0.1	0.3	1	3	10
8.0, 8.0	702	703	706	715	752	866	865	876	873
8.0, 4.0	177	178	178	179	183	197	221	212	205
8.0, 2.0	46	46	46	46	46	48	56	58	55
8.0, 1.0	12	12	12	12	12	13	14	19	18
8.0, 0.5	4	4	4	4	4	4	4	6	5
8.0, 0.25	1	1	1	1	1	1	1	2	4
8.0, 0.125	1	1	1	1	1	1	1	1	2
8.0, 0.0	1	1	1	1	1	1	1	1	1
4.0, 8.0	687	687	685	689	739	803	827	838	841
4.0, 4.0	174	174	174	175	182	207	205	203	198
4.0, 2.0	45	45	45	45	47	52	59	55	53
4.0, 1.0	12	12	12	12	13	14	19	19	18
4.0, 0.5	3	3	3	4	4	4	6	7	7
4.0, 0.25	1	1	1	1	1	1	2	4	4
4.0, 0.125	1	1	1	1	1	1	1	2	2
4.0, 0.0	1	1	1	1	1	1	1	1	1
2.0, 8.0	1006	983	927	860	808	758	786	839	826
2.0, 4.0	256	254	247	235	228	209	197	198	192
2.0, 2.0	66	66	65	64	64	64	57	56	56
2.0, 1.0	18	18	18	18	19	21	20	19	19
2.0, 0.5	5	5	5	5	6	7	8	8	8
2.0, 0.25	2	2	2	2	2	3	4	4	4
2.0, 0.125	1	1	1	1	1	1	2	2	2
2.0, 0.0	1	1	1	1	1	1	1	1	1
1.0, 8.0	1700	1618	1389	1048	787	721	758	835	826
1.0, 4.0	434	430	397	324	244	202	191	196	193
1.0, 2.0	113	113	111	102	80	66	58	55	54
1.0, 1.0	30	31	31	31	28	24	20	20	19
1.0, 0.5	9	9	10	10	10	9	8	8	8
1.0, 0.25	3	3	4	4	4	4	4	4	4
1.0, 0.125	2	2	2	2	2	2	2	2	2
1.0, 0.0	1	1	1	1	1	1	1	1	1
0.5, 8.0	1703	1655	1456	1147	831	705	738	821	826
0.5, 4.0	433	432	409	350	263	205	193	190	193
0.5, 2.0	112	113	112	104	85	67	59	55	53
0.5, 1.0	31	31	31	31	28	25	20	19	19
0.5, 0.5	10	10	10	10	10	9	9	8	8
0.5, 0.25	4	4	4	4	4	4	4	4	4
0.5, 0.125	2	2	2	2	2	2	2	2	2
0.5, 0.0	1	1	1	1	1	1	1	1	1
0.25, 8.0	1698	1655	1523	1181	828	730	742	814	826
0.25, 4.0	434	431	414	376	280	209	192	187	194
0.25, 2.0	112	113	112	107	94	73	57	54	53
0.25, 1.0	31	31	31	31	29	26	21	20	19
0.25, 0.5	10	10	10	10	10	9	9	8	8
0.25, 0.25	4	4	4	4	4	4	4	4	4
0.25, 0.125	2	2	2	2	2	2	2	2	2
0.25, 0.0	1	1	1	1	1	1	1	1	1
0.125, 8.0	1705	1692	1638	1291	869	713	743	810	826
0.125, 4.0	433	433	429	385	292	211	188	187	194
0.125, 2.0	112	112	113	112	97	74	59	53	53
0.125, 1.0	31	31	31	31	31	27	22	19	19
0.125, 0.5	10	10	10	10	10	10	9	8	8
0.125, 0.25	4	4	4	4	4	4	4	4	4
0.125, 0.125	2	2	2	2	2	2	2	2	2
0.125, 0.0	1	1	1	1	1	1	1	1	1
0.0, 8.0	1681	1681	1655	1435	862	692	745	807	826
0.0, 4.0	425	425	425	424	310	215	184	187	194
0.0, 2.0	111	111	111	111	100	74	61	54	52
0.0, 1.0	31	31	31	31	31	29	22	19	19
0.0, 0.5	10	10	10	10	10	10	9	9	8
0.0, 0.25	4	4	4	4	4	4	4	4	4
0.0, 0.125	2	2	2	2	2	2	2	2	2
0.0, 0.0	1	1	1	1	1	1	1	1	1

To use the force limits in Tables 1-3, both the residual and modal masses of the source and load must be known as a function of frequency, either from a test, from an FEM, or from both. FEM analyses provide both the modal and residual effective masses (see Section 3.3.4). If shaker or tap tests are used to measure the effective masses, the smoothed FRF of the magnitude of the ratio of force to acceleration should be taken as the effective residual mass, as discussed in Section 3.3.3.

4.2 Semi-empirical Method of Predicting Force Limits

4.2.1 Rationale

The semi-empirical approach to deriving force limits is based on the extrapolation of interface force data for similar mounting structure and test items. The following form of semi-empirical force limit for sine or transient tests was proposed in 1964 [3]:

$$F_l = C \ M_o \ A_s \qquad (27)$$

where F_l is the amplitude of the force limit, C is a frequency dependent constant which depends on the configuration, M_o is the total mass of the load (test item), and A_s is the amplitude of the acceleration specification. The form of Eq. 27 appropriate for random vibration tests is:

$$S_{FF} = C^2 \ M_o^2 \ S_{AA} \qquad (28)$$

where S_{FF} is the force spectral density and S_{AA} the acceleration spectral density.

In [3], it is claimed that C seldom exceeds 1.4 in coupled systems of practical interest, because of the vibration absorber effect. From the preceding analysis of TDFS's, it is apparent that this claim implies something about the ratio of the load and source effective masses of the coupled system. For the simple TDFS shown in Fig. 2, the ordinate of Fig. 5 may be interpreted as C^2 in Eq. 28. Thus a value of C of 1.4 corresponds to an ordinate value of 1.96 and an abscissa value, the ratio of load to source mass (M_2/M_1), of 1.5. Although the load and source effective masses are often comparable in aerospace structures, such is not always the case, and the use of Eqs. 27 or 28 with $C = 1.4$ might result in undertesting for lightweight loads mounted on heavy structure.

A refinement of Eq. 28 follows from inspection of Eq. 19, which is Newton's second law for random vibration. If one takes the envelope of both sides of Eq. 19, the left-hand-side is the envelope of the interface force spectrum, which is the sought-after force limit. On the right-hand-side results the envelope of the product of the load apparent mass and the interface acceleration spectrum. This product may be approximated as the frequency average (the aforementioned asymptote) of the magnitude squared of the load apparent mass, times the envelope of the interface acceleration spectrum. This refinement of the semi-empirical approach was the subject of several important papers in the 1970's [10,13 & 14]. Herein, it will be assumed that the asymptote of the magnitude of the load apparent mass is equal to the total mass below the first resonance, and then falls off as one over frequency at frequencies above the first resonance, as is the case for the rod example in Fig. 4 and for a plate excited in bending. (However, the asymptotic mass of a beam excited in bending, falls off as one over the square root of frequency.) Assuming a one-over-frequency fall-off of the asymptotic load mass, leads to the following modification of Eq. 28, applicable to random vibration testing:

$$S_{FF} = C^2 \, M_o^2 \, S_{AA}, \qquad\qquad f < f_o$$

$$\text{(29)}$$

$$S_{FF} = C^2 \, M_o^2 \, S_{AA} \, / \, (f \, / \, f_o)^2 \qquad f > f_o$$

Some judgment and reference to test data for similar configurations must be considered to choose the value of C and the exponent of (f/f_o) Eq. 29. Herein are considered force data measured at the interface between three equipment items and the Cassini spacecraft Development Test Model (DTM) during acoustic tests.

4.2.2 Validation

Figure 7 shows a schematic of the Cassini spacecraft which will be launched in October 1997 to arrive at Saturn in 2004. Figure 8 shows the DTM spacecraft configured for one of several acoustic tests in the JPL reverberant acoustic chamber [39]. The three equipment items which were instrumented with tri-axial force transducers between their mounting structure and the spacecraft ring-stringer structures are: a dynamic model of a Radioisotope Thermoelectric Generator (RTG) three of which provide the spacecraft electrical power, the engineering model Radio Plasma Wave Subsystem (RPWS), and the flight Propulsion Module Subsystem Electronic Assembly (PMSEA). The RTG weighted approximately 120 lb. and was cantilevered outward from a mounting bracket attached at four points to the spacecraft lower equipment module. The RPWS weighed approximately 65 lb. and was attached with three trusses to the spacecraft upper equipment module. The PMSEA is a large electronic box which also weighed approximately 65 lb. and was mounted at four corners to the propulsion module.

Figure 9 shows the spectra of the total vibration force acting in three directions at the PMSEA/spacecraft interface during the DTM protoflight level acoustic test. Figure 10 shows the acceleration spectra measured at the PMSEA/spacecraft interface in the acoustic test. Figure 11 shows the flight PMSEA mounted on a shaker for a vertical vibration test. Figure 12 shows the magnitude (reduced by 4) of the PMSEA apparent mass measured in a preliminary low-level (0.25 G) sine sweep vertical axis vibration test. Notice that the fundamental resonance of the PMSEA in the shaker vertical test is at approximately 400 Hz, whereas the fundamental (radial) resonance of the PMSEA mounted on the spacecraft DTM (Fig. 9) is approximately 100 Hz. This disparity between the fundamental resonance frequencies on the flight structure and on the shaker is typical, and it should discourage anyone from thinking that the shaker test is a close replica of the field environment. The goal of force control is simply to limit the maximum force on the shaker, to that estimated for flight, and it must be recognized that the frequencies at which this maximum force is exhibited will be different in the two configurations.

Also shown in Fig. 9 are the semi-empirical vertical force specification calculated from Eq. 29 and the force limit used in the vertical vibration test. In the case of the PMSEA, the vibration test force specification was based on an envelope of the DTM acoustic test measurements. (It is very unusual to have system acoustic test data available before the instrument vibration test.) The semi-empirical force specification is based on the acceleration spectrum in Fig. 10 and a value of C of unity. Because the acceleration specification in Fig. 10 drops at 250 Hz, the semi-empirical force limit is about 7 dB less than the vibration test force limit at 400 Hz. The semi-empirical force limit is a reasonable envelope of the acoustic test force data, particularly in the mid-frequency range where structure-borne random vibration is the dominant source. It should be noted that the acoustic test data as well as the semi-empirical force limit, which is proportional to the acceleration specification, have a 4 dB margin over the predicted flight environment. Fig. 13 shows the notching, approximately 10 dB, that resulted in the PMSEA vertical random vibration test, when the shaker force was limited to the vibration test specification in Fig. 9.

Figure 14 compares the semi-empirical and vibration test vertical force specifications for the Cassini RPWS instrument with the interface force data measured in the DTM acoustic test. Both force specifications envelope the interface force data peaks at approximately 45 Hz in the radial direction and at 65 Hz in the lateral directions. Figure 15 shows the interface acceleration data measured at the RPWS/DTM interface. One of the lateral acceleration measurements greatly exceeds the specification. Figure 16 shows the RPWS configured for a lateral vibration test. Figure 17 shows the magnitude (reduced by 4) of the RPWS apparent mass measured in a preliminary low-level (0.25 G) sine sweep vertical axis vibration test. Notice that the fundamental resonance of the RPWS in the shaker vertical test is at approximately 250 Hz, whereas the fundamental (radial) resonance of the RPWS mounted on the spacecraft DTM (Fig. 14) is approximately 45 Hz. This discrepancy explains why the semi-empirical force specification does not roll off until the first resonance frequency on the shaker and why the force specifications in Fig. 14 must greatly exceed the DTM data in the high frequency regime. Figure 18 shows the notching that resulted in the full-level vertical random vibration test of the RPWS, when the test force specification in Fig. 15 was utilized.

Figures 19 and 20 show the Cassini RTG interface forces and accelerations measured at the RTG/DTM interface. The acceleration test specification in Fig. 20 is a very accurate envelope of the DTM acoustic test data. The first resonance in the radial direction on the spacecraft DTM is at approximately 220 Hz. (The radial direction on the spacecraft is along the RTG axis.) The semi-empirical force limit in Fig. 19 is flat to 750 Hz because the first RTG resonance in the vertical vibration test shown in Fig. 21 was at approximately 750 Hz. The RTG's were inherited hardware which had been previously qualified to an existing test specification. The force specification in Fig. 19 was used to justify the extension of the previous qualification testing to the Cassini program.

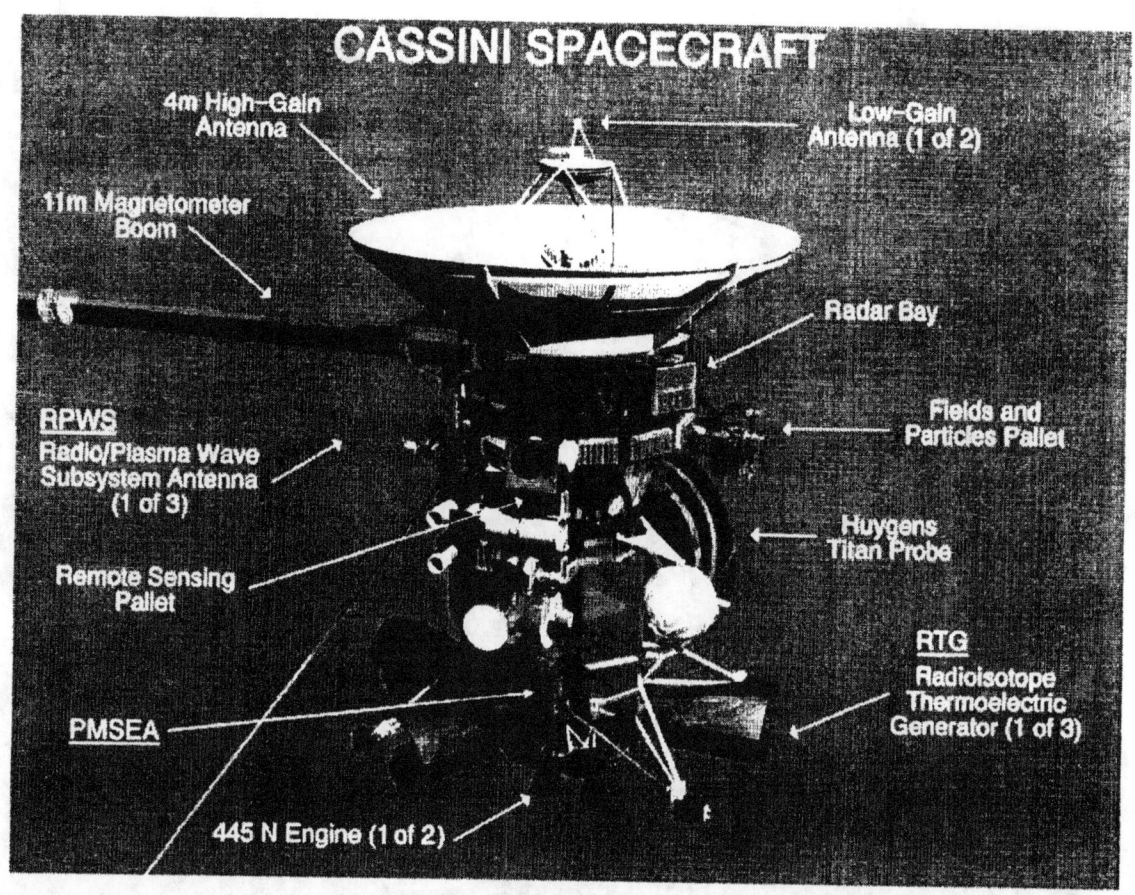

FIGURE 7. Cassini Spacecraft Schematic (Showing subject PMSEA, RPWS, and RTG's)

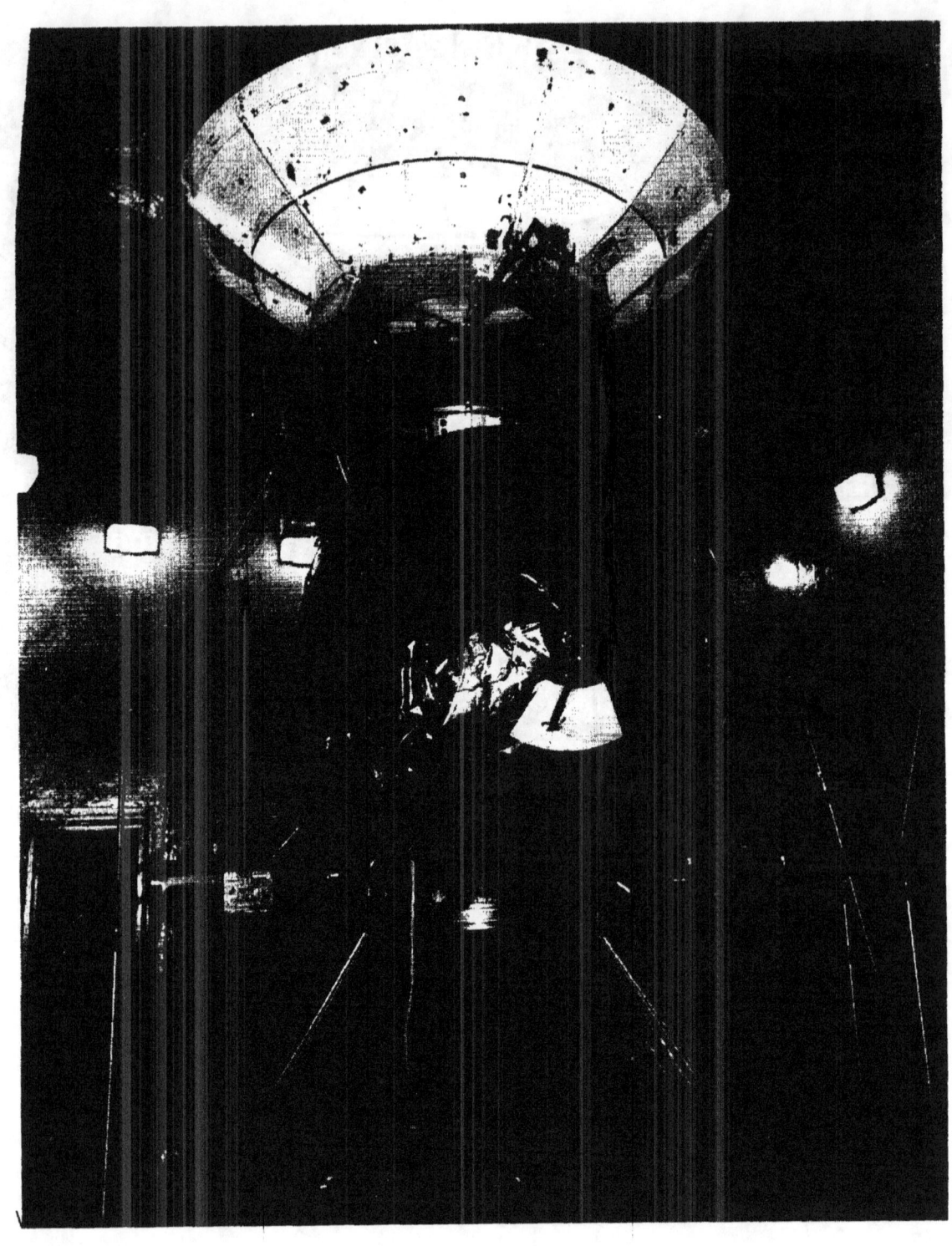

FIGURE 8. Cassini DTM Spacecraft in Acoustic Chamber

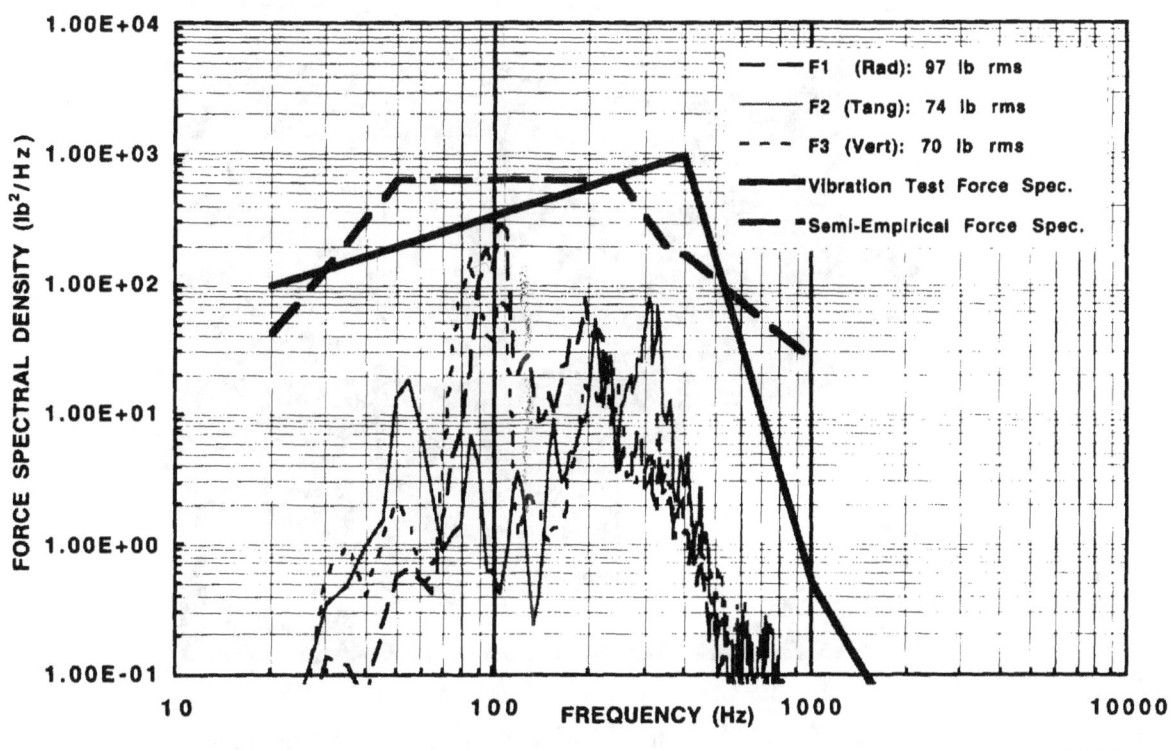

FIGURE 9. Comparison of Specified and Measured PMSEA Interface Forces

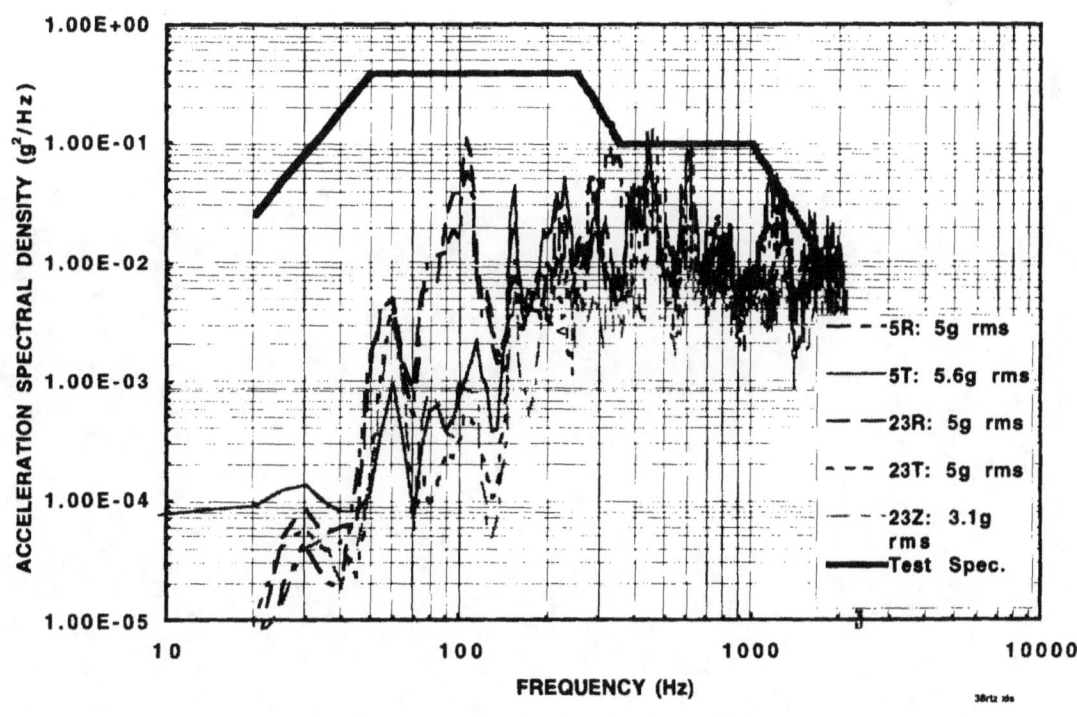

FIGURE 10. Comparison of Specified and Measured PMSEA Interface Accelerations

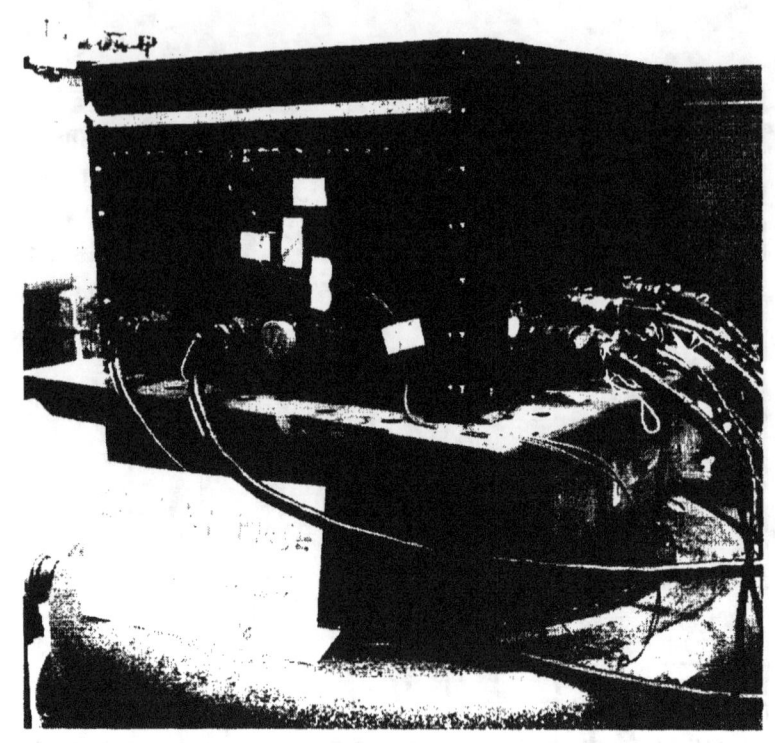

FIGURE 11. Cassini PMSEA In Vertical Vibration Test

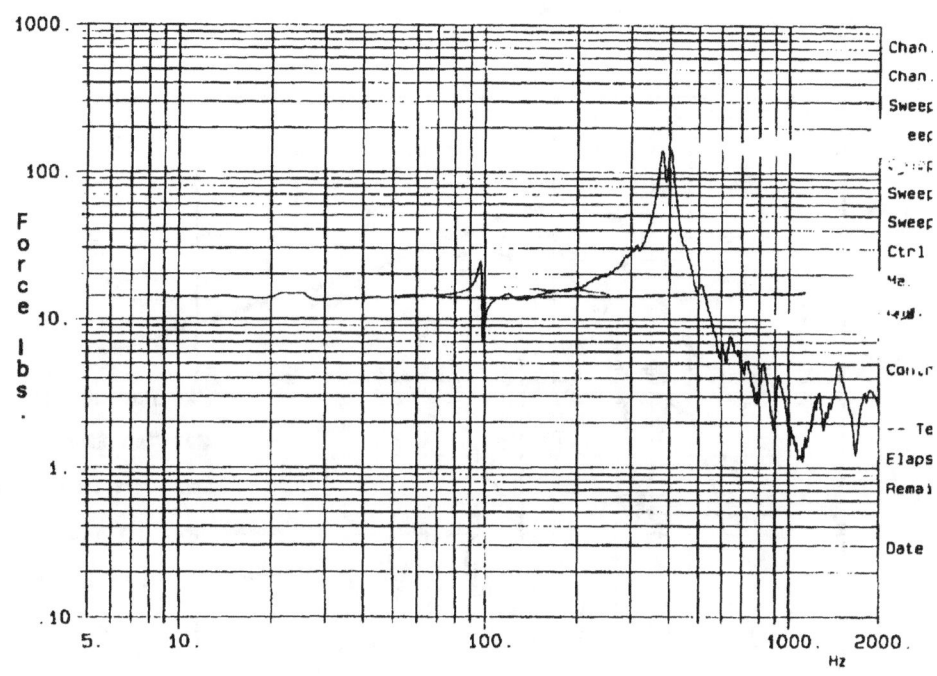

FIGURE 12. Force in 1/4 G Sine-sweep Vertical Vibration Test of Cassini PMSEA

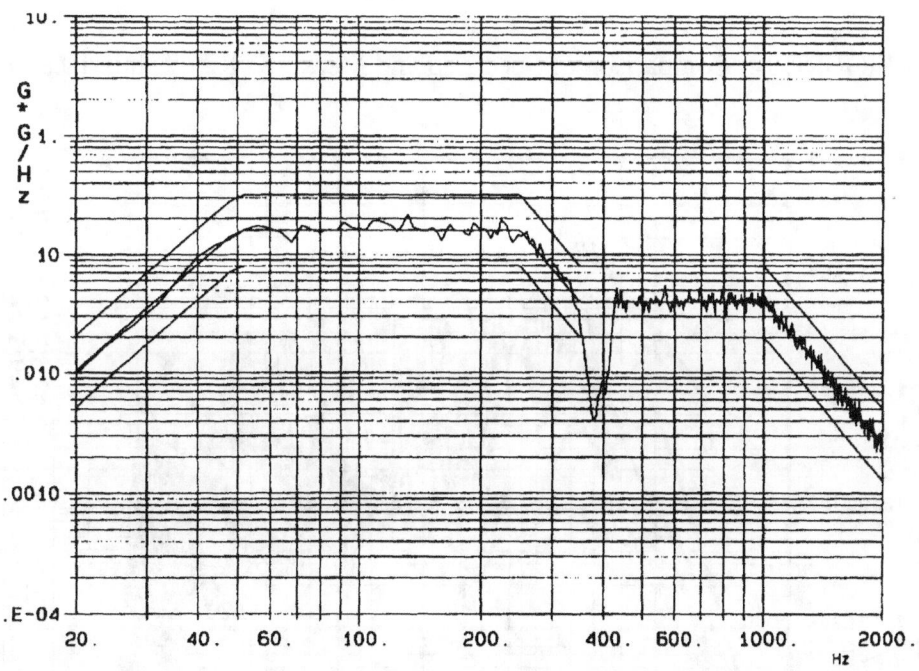

FIGURE 13. Notched Acceleration Input in Vertical Random Vibration Test of PMSEA

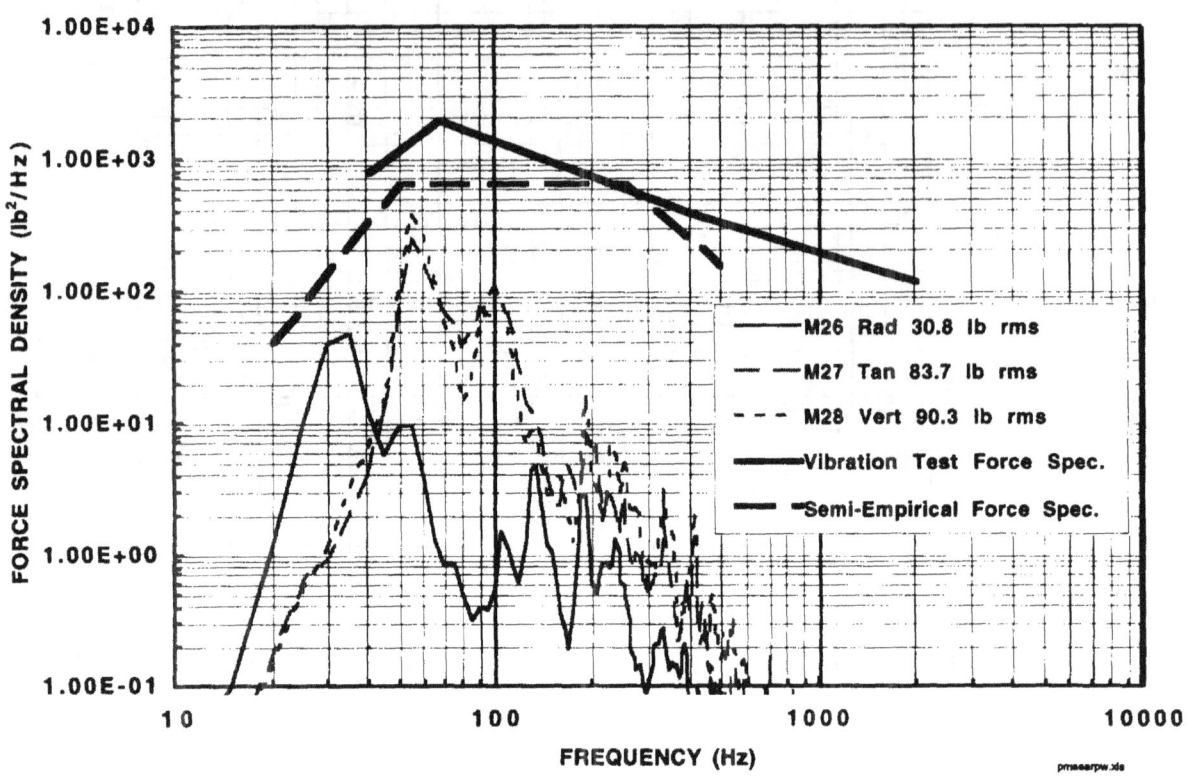

FIGURE 14. Comparison of Specified and Measured RPWS Interface Forces

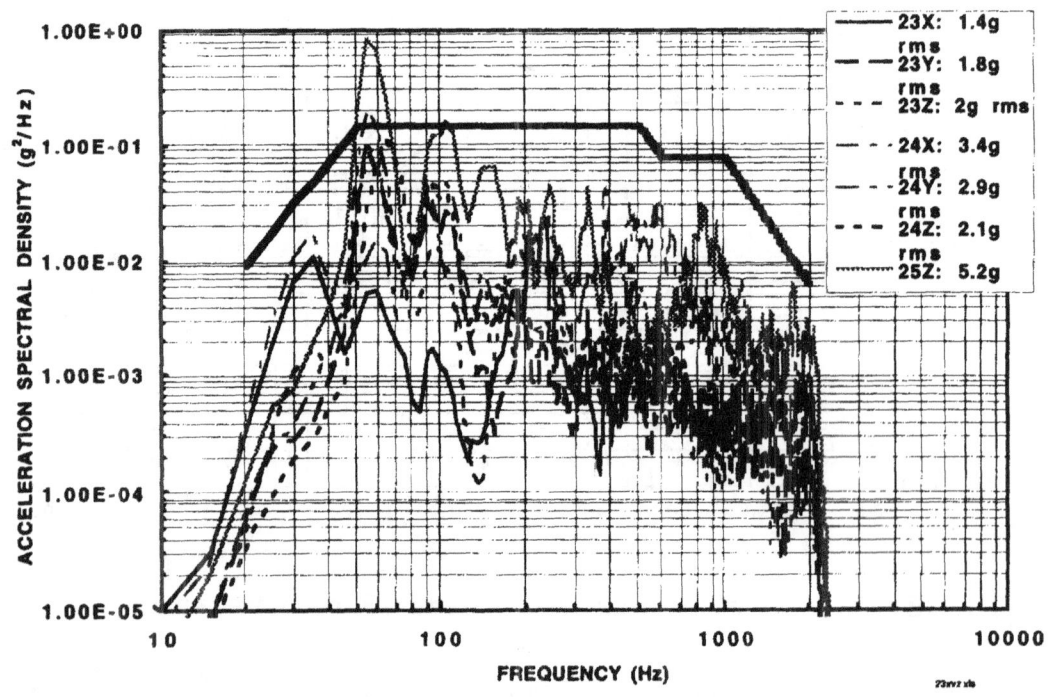

FIGURE 15. Comparison of Specified and Measured RPWS Interface Accelerations

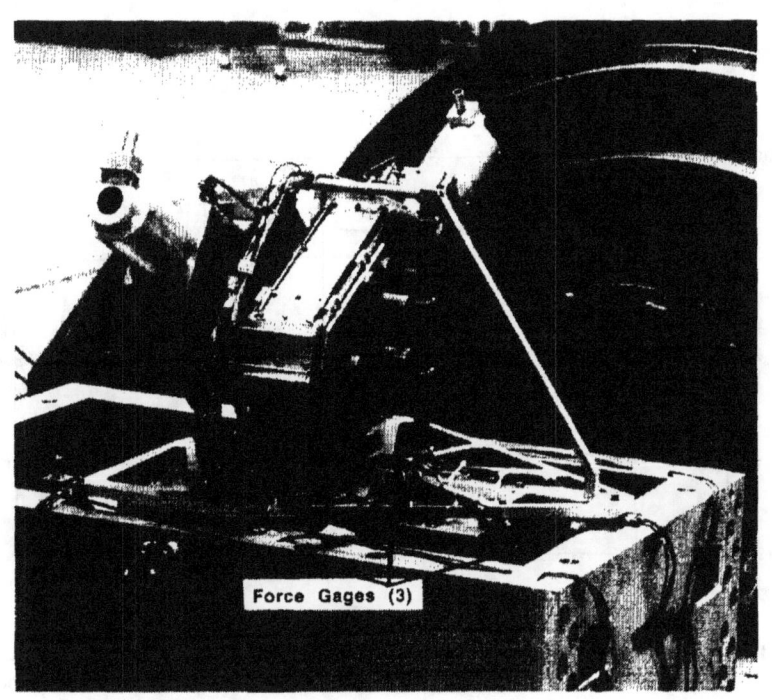

FIGURE 16. Cassini RPWS In Lateral Vibration Test

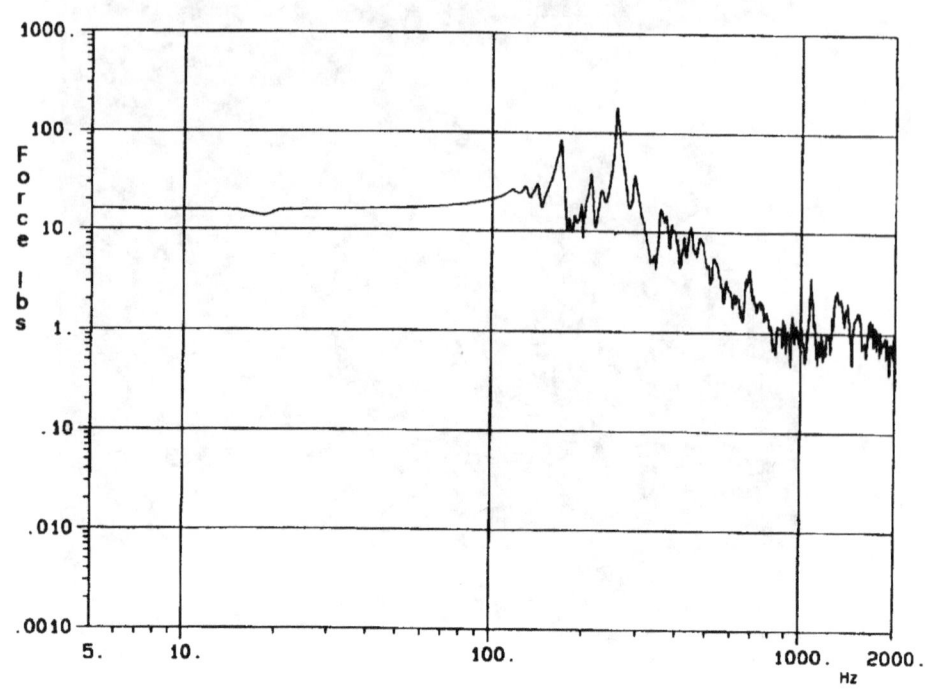

FIGURE 17. Force in 1/4 G Sine-sweep Vertical Vibration Test of Cassini RPWS

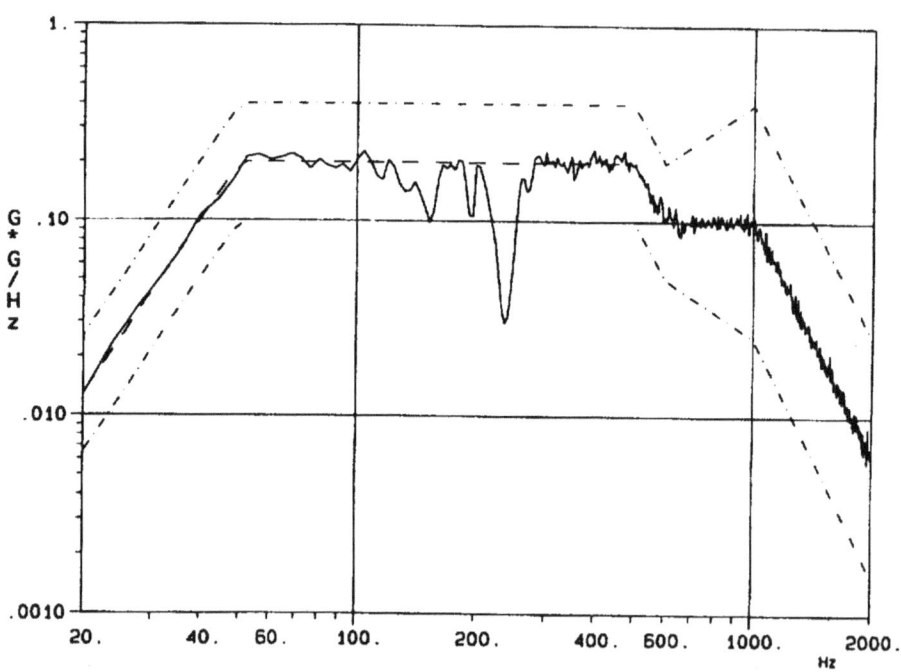

FIGURE 18. Notched Acceleration Input in Vertical Random Vibration Test of RPWS

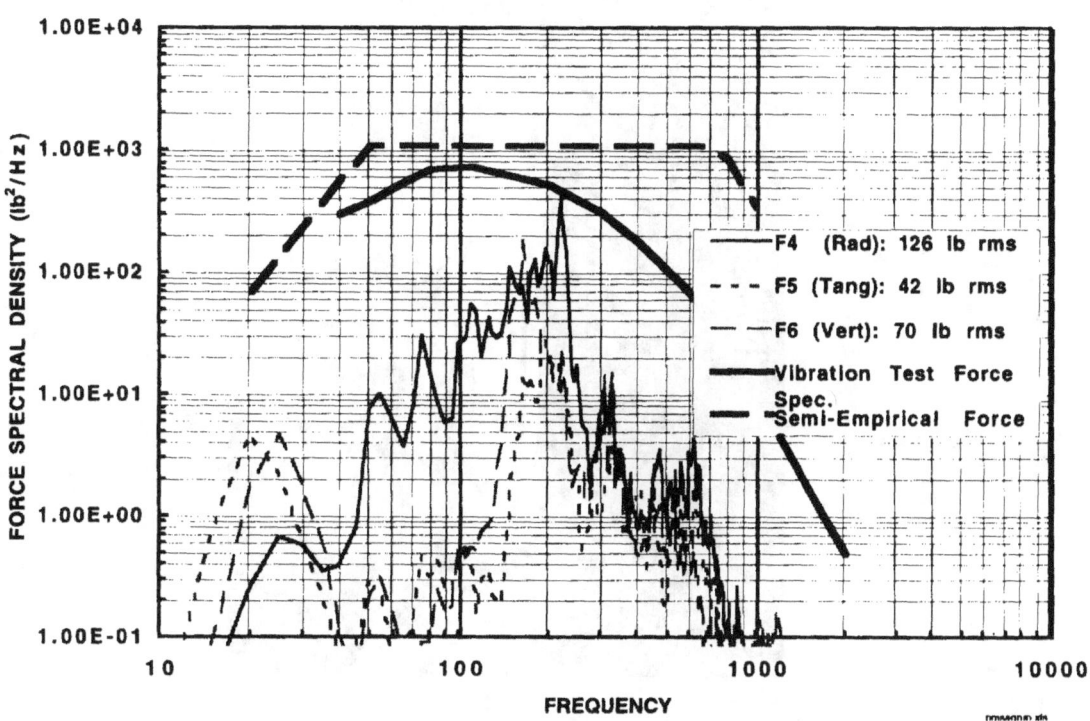

FIGURE 19. Comparison of Specified and Measured RTG Interface Forces

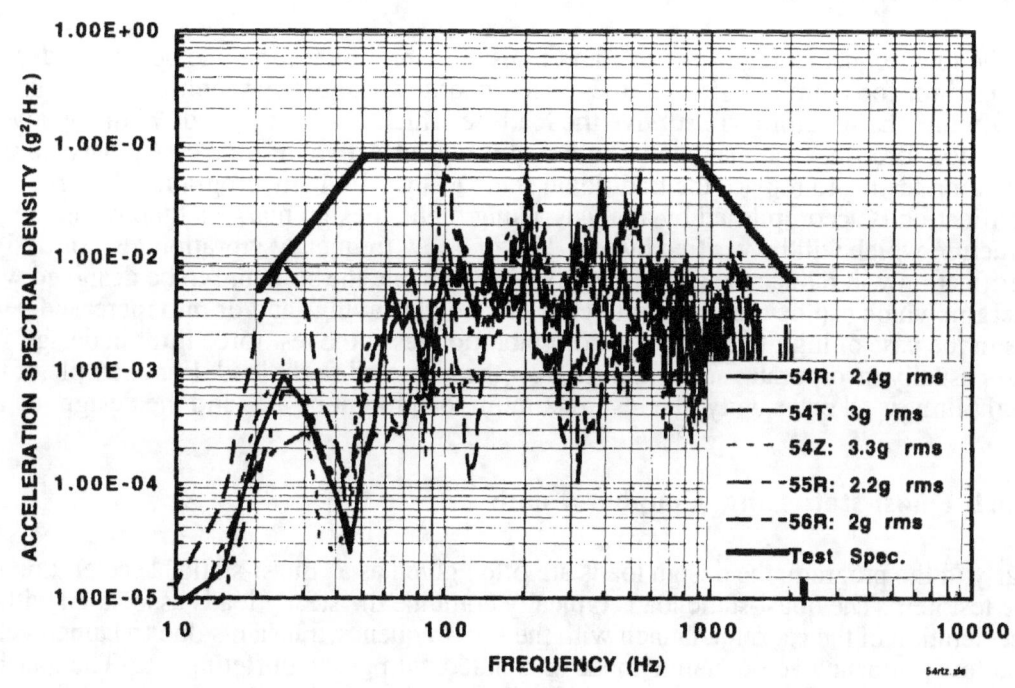

FIGURE 20. Comparison of Specified and Measured RTG Interface Accelerations

FIGURE 21. Cassini RTG In Vertical Vibration Test

4.3 Design Load Verification

In the low frequency regime of sine and transient vibration testing, the design loads may often be utilized for force limiting. In fact, the objective of low frequency vibration testing is usually to verify that the structure will survive the loads to which it was designed. With the advent of vibration force measurement and limiting, it is cost effective in some cases to use low frequency vibration testing to replace the traditional static testing. In many programs, the structural design verification is accomplished by analysis, using higher design margins than would be used for a structure which will be verified by test. Use of a low frequency vibration test for design verification will reduce the costs of analysis and enable the structure to be designed with less margin, which can be translated into cost and weight savings and/or into increased performance. As in the case of high frequency random vibration tests, the test force limit in design verification tests usually incorporates some margin over the expected flight load. For example, if the expected flight limit is taken as unity, the test maximum load might be 1.2, and the design load 1.5.

4.3.1 Quasi-static and Coupled Loads

Early in the program, the design loads are often given as a "quasi-static" acceleration of the CG of the test item. The quasi-static loads typically combine the steady loads associated with the acceleration of the rocket at launch with the low frequency transients due to launch vehicle staging and low frequency aerodynamic loads associated with gusts, buffeting, etc. The quasi-static loads are interpreted by the designer as static loads. Therefore these quasi-static accelerations are multiplied by the total mass of the test item to obtain the applied loads used for preliminary design, for selecting the fasteners, etc. These quasi-static accelerations are obtained from past experience or from semi-empirical curves developed for different launch vehicles and spacecraft configurations [40].

Later in the program, the low frequency loads used for design and for analytical verification of the designs are refined by analysis of coupled finite element models (FEM's) [41]. The spacecraft loads are determined from a coupled spacecraft and launch vehicle model, and later the equipment loads are determined from a coupled equipment and launch vehicle model. Obviously the validity of the FEM models depends on the skill of the modeler and upon the accuracy of the information used to develop the model. More sophisticated models are used as the programs mature. For example, in preliminary analyses of a spacecraft, much of the equipment may be modeled as a lumped masses at the equipment CG's. In this case, the calculated forces at the spacecraft and equipment interfaces are only valid at frequencies below the first resonance frequency of the equipment. Later in the program, the equipment may be represented by complex FEM's. But even these models are usually valid only in the frequency regime encompassing the first few modes of the equipment in each axis. One objective of coupled loads analyses is to maximize the upper frequency limit of the model. The validity of the models is often verified and extended to higher frequencies by refining the models with modal test data.

4.3.2 Force Transducer Measurement of CG Acceleration

Although it is relatively easy to predict the CG acceleration, it is difficult or impossible to measure the acceleration of the CG with accelerometers in a vibration test. Sometimes the CG is inaccessible, or there is no physical structure at the CG location on which to mount an accelerometer. However, there is a more serious problem. Only in the case of a rigid body is the CG a fixed point on the structure. Once deformations and resonances and occur, it is impossible to measure the CG acceleration with an accelerometer. Unfortunately, attempts to measure the CG acceleration with an accelerometer usually overestimate the CG response at resonances, so limiting these measurements to the CG criterion will result in an undertest. However, the CG acceleration is uniquely determined by dividing an interface force measurement by the total mass of the test item, per Newton's second law.

The non-fixity of the CG of a deformable body is demonstrated with an example in Fig. 22, which illustrates the third vibration mode of a three-mass, two-spring vibratory system. The mass value of the middle mass is twice that of the end masses and the two springs are identical. The upper sketch in Fig. 22 shows the system at rest with the CG clearly located at the center of the middle mass. The lower sketch in Fig. 22 shows the system displaced in its third mode, with the middle mass moving one unit to the left and the two end masses both moving one unit to the right. (The first mode involves rigid body translation, and the second mode involves zero motion of the middle mass and the two end masses moving an equal amount in opposite directions.) It is a characteristic of modal motion that there are no external forces acting, so by Newton's second law, the modal displacement illustrated in Fig. 22 can't involve motion of the CG. However, since the middle mass, as well as the end masses, are moving, the CG is not at a fixed point in the system. Clearly, one could not attach an accelerometer at the CG position.

The difficulty of measuring the CG acceleration with an accelerometer is further illustrated with data obtained on the RPWS instrument (Fig. 16) in the Cassini spacecraft DTM acoustic test (Fig. 8). Figure 23 is a schematic of the Cassini RPWS instrument shown in Fig. 16. In the Cassini DTM spacecraft acoustic test, the RPWS was instrumented with tri-axial force transducers between the instrument and spacecraft. (See force data in Fig. 14.) In addition to the interface accelerometers, there was also a tri-axial accelerometer located approximately at the CG of the RPWS instrument, in the spacecraft DTM acoustic test. (See position 23 in Fig. 23.) Fig. 24

shows the ratio of the total external radial force to the radial CG acceleration for the RPWS in the DTM spacecraft acoustic test, and Fig. 25 shows the corresponding ratio for a lateral axis. If the tri-axial accelerometer actually measured the CG acceleration, the curves in both Figs. 24 and 25 would be flat functions of frequency equal to the total weight of the RPWS, approximately 65 lb. In the spacecraft radial direction Fig. 24 shows that the ratio of force to acceleration falls off rapidly above about 160 Hz. (Fig. 17 shows that there is an RPWS resonance at approximately 160 Hz in the RPWS vertical direction which corresponds to the spacecraft radial direction.) In the spacecraft lateral direction, Fig. 25 shows that the ratio falls off above 50 Hz, which corresponds to a lateral resonance of the RPWS on the spacecraft. That the measured ratios of force to acceleration are less than the total weight indicates that the measured accelerations are greater than the true CG acceleration at the higher frequencies.

The examples in Figs. 22-25 demonstrate the problems and dangers of using an accelerometer to measure acceleration of the CG in vibration tests. It is for this reason that CG response limiting and quasi-static design verification have been difficult, at best, to implement previously in vibration tests. However, with the advent of piezo-electric, tri-axial force transducers, these measurements become straight-forward, and the use of vibration tests for design verification becomes very attractive.

4.4 Force Limiting Vibration Example--ACE CRIS Instrument

Figure 26 is a photograph of the ACE spacecraft Cosmic Ray Isotope Spectrometer (CRIS) instrument mounted on a shaker for a vertical vibration test. The instrument is mounted on twelve uni-axial force transducers, which stay with the instrument in flight and record the lift-off vibratory forces.

Figure 27 is a spreadsheet for calculating the force limits three ways: the simple TDFS method, the complex TDFS method, and the semi-empirical method. The spreadsheet is linked to Eqs. 20 and 21 for the simple TDFS calculation and to Tables 1-3, plus an interpolation routine, for the complex TDFS calculation. A value of Q of 50, 20 or 5, corresponding to Tables 1, 2, and 3, must be chosen for the complex TDFS calculation. The force limits are calculated in one-third octave bands from 40 to 1000 Hz. (This frequency range is typical for an instrument, but might include lower frequencies in the case of a spacecraft test.) There is nothing sacred about one-third octave bands. One could choose one-tenth or alternately, octave bands, since specifications are relatively smooth functions of frequency. (It should be noted that the width of the notching is set by the shaker controller analysis bandwidth, typically 5 Hz for random vibration tests, not by the specification bandwidth.) The input acceleration specification must be entered for each one-third octave band. (Recall that the force limit is proportional to the acceleration specification for all three methods.) The remainder of the inputs in the spreadsheet deal with the structural impedance characteristics of the load (the test item) and of the source (the mounting structure). For both the load and source, one must enter the residual weight and the number of modes in each one-third octave band. The spreadsheet calculates modal weight, and then the force limits are automatically calculated using the three aforementioned methods. The residual weight information may be determined from test, FEM, or a combination of these. In this example the information was determined from test data.

Figure 28 shows the magnitude of the apparent mass of an ACE spacecraft honeycomb panel measured in a tap test at one of the CRIS instrument mounting locations. (This type of test involves tapping at the selected point on the panel near an accelerometer with a small hammer which incorporates a force transducer, and using a two channel frequency analyzer to compute the magnitude of the apparent mass.) The apparent mass data are smoothed in frequency to compute the asymptotic value, which is taken as the residual mass, as discussed in section 3.3.3. The asymptotic value in Fig. 28 rolls off as one over frequency squared above 50 Hz, which is characteristic of a spring. Below 50 Hz, the apparent mass looks like a mass, but it is known that the coherence fell off below 50 Hz, so the data below 50 Hz are suspect. To obtain the residual mass of the source for the spreadsheet, the data in Fig. 28 was multiplied by four to approximately account for the multiple (twelve) mounting points. (At low frequencies, the residual mass should approach the total mass or stiffness.) In the Fig. 27 spreadsheet, it is assumed that there is one source mode in every one-third octave band, so the decrease in the residual mass in each band becomes the average modal mass in that band. (When FEM information is used, the number of significant modes in each one-third octave band are counted.)

Figure 29 shows the magnitude of the apparent mass of the CRIS instrument measured on the shaker in Fig. 26. The apparent mass has the characteristics of a vibrating plate or rod driven at a point, i.e. it is equal to the total mass below the first resonance frequency and then rolls off like one over frequency. (See section 3.3.3.) The asymptotic values in Fig. 29 are multiplied by four to scale to one G of input, and entered as the load residual mass in the spreadsheet. (Notice that the roll-off of the asymptotic mass is started one-third octave band below the resonance frequency in Fig. 29 to generate some modal mass at the first resonance frequency.) It is also assumed that there is one mode of the load in every one-third octave band above the first resonance frequency.

The plot in Fig. 27 shows the force limit calculated with each of the three methods, using a constant $C = 1.5$ in Eq. 29 for the semi-empirical method. In this case, the semi-empirical method approximately splits the difference between the two TDFS methods, with the simple TDFS method giving higher limits below the first resonance and the complex TDFS method giving higher limits above the first resonance.

Figure 30 shows the total vertical force limited to a force specification in the protoflight vertical random vibration test of the CRIS instrument shown in Fig. 26. The force specification actually used in the test, which is shown in Fig. 30, is very similar to that predicted with the semi-empirical method in the spreadsheet of Fig. 27. Figure 31 shows the notch in the acceleration input which resulted from the force limiting in Fig. 30. Comparison of force limited and unlimited data from low level runs indicates that the notch in the acceleration spectrum which results from force limiting is typically the mirror image of that portion of the force spectrum which would have exceeded the force limit, if no limiting were implemented. Notice the asymmetric shape of the notch in Fig. 31. It is characteristic that the notch, as well as the unlimited force, are steeper functions of frequency above a resonance than below it. This is in contrast to manual notches which are usually designed to be symmetric.

The ACE spacecraft instrumentation includes a Spacecraft Launch Acceleration Measurement (SLAM) data acquisition system to measure, record, and transmit dynamic data at launch of the spacecraft on a Delta launch vehicle. The SLAM instrumentation includes a channel for the high frequency (20 to 2000 Hz) acceleration measured normal to the spacecraft honeycomb panel near one of the twelve mounting feet of the CRIS instrument and also a channel for the total normal force measured under the twelve mounting feet of the CRIS instrument. (The CRIS instrument is mounted on twelve uni-axial force transducers, and the output of these transducers is summed.) Figs. 32 and 33 show the spectral densities of these acceleration and force data measured during the prototype level acoustic test of the ACE spacecraft at NASA Goddard Space Flight Center (GSFC). The CRIS acceleration spectrum in Fig. 32 measured in the spacecraft acoustic test is approximately an order of magnitude less than the CRIS random vibration test acceleration input spectrum in Fig. 31. This is unusually conservative, particularly when it is considered that the acoustic test spectrum is probably a conservative envelope of the flight acoustic levels. The CRIS normal force spectrum in Fig. 33 measured in the spacecraft acoustic test is approximately two orders of magnitude less than the CRIS random vibration test force limit spectrum in Fig. 30. One order of magnitude of conservatism in the force specification can be attributed to the conservatism of the acceleration spectrum, since the force specification is proportional to the acceleration specification. However, the other order of magnitude of conservatism in the force spectrum must be attributed to the methods used to derive the force spectrum. It must be concluded that, even with force limiting, the random vibration test of the CRIS instrument was a severe overtest.

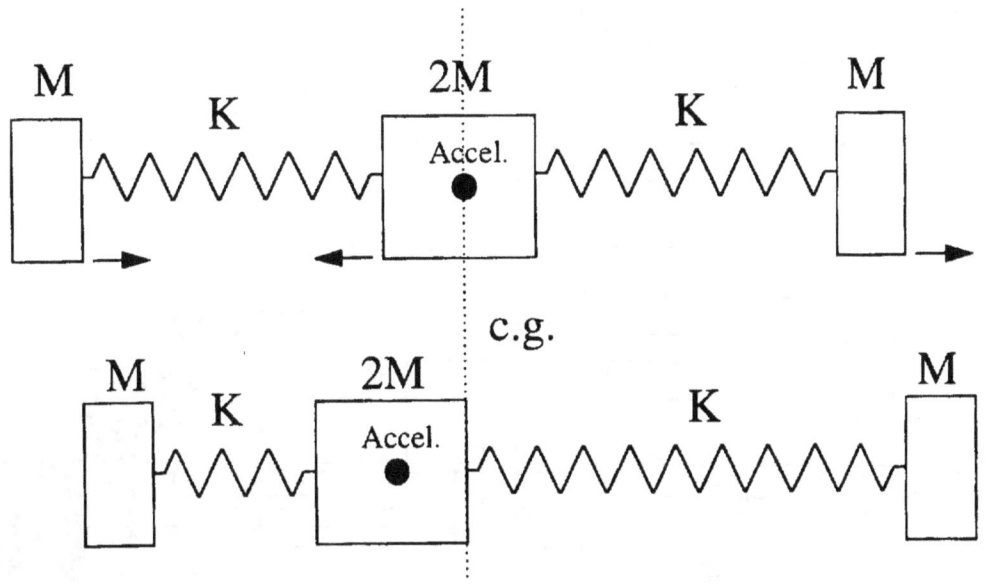

FIGURE 22. Third Mode of Three-Mass, Two-Spring System Showing Non-fixity of CG

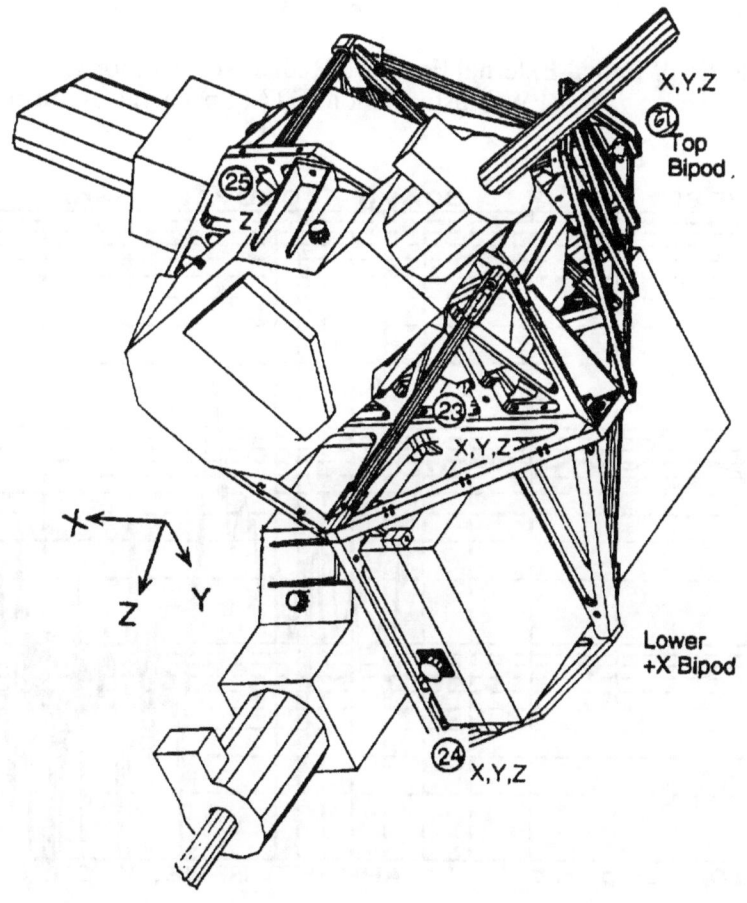

FIGURE 23 Schematic of Cassini RPWS Instrument Showing Accelerometers

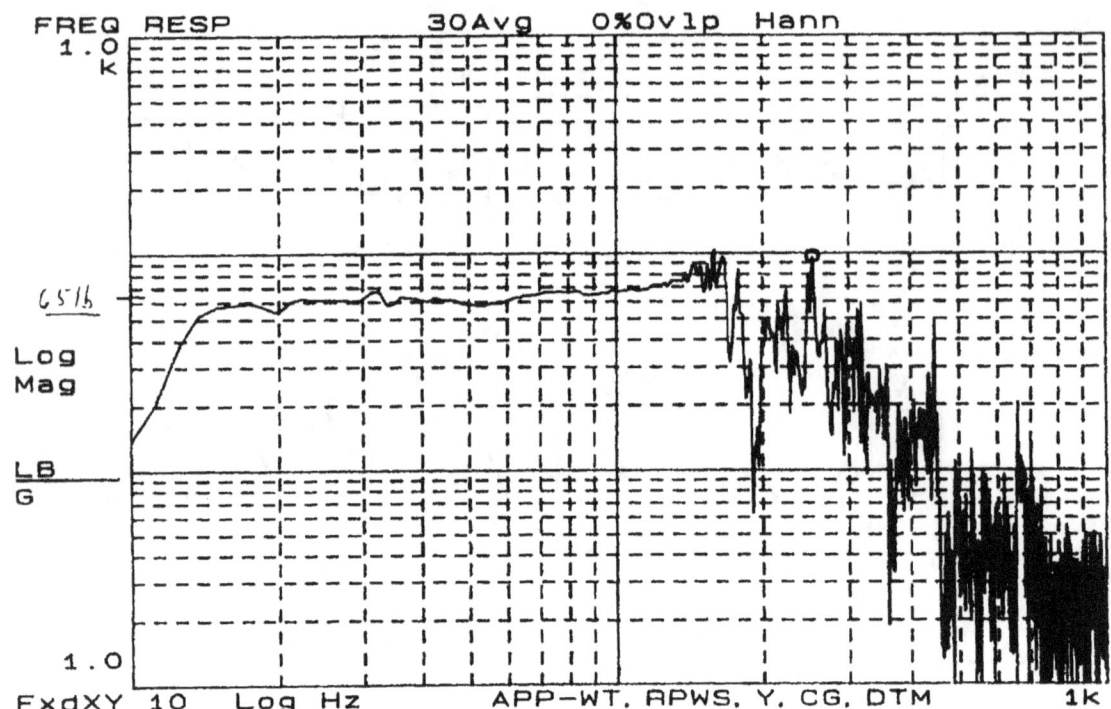

FIGURE 24. Ratio of Radial External Force To Radial Acceleration at Nominal CG for Cassini
RPWS Instrument in DTM Spacecraft Acoustic Test

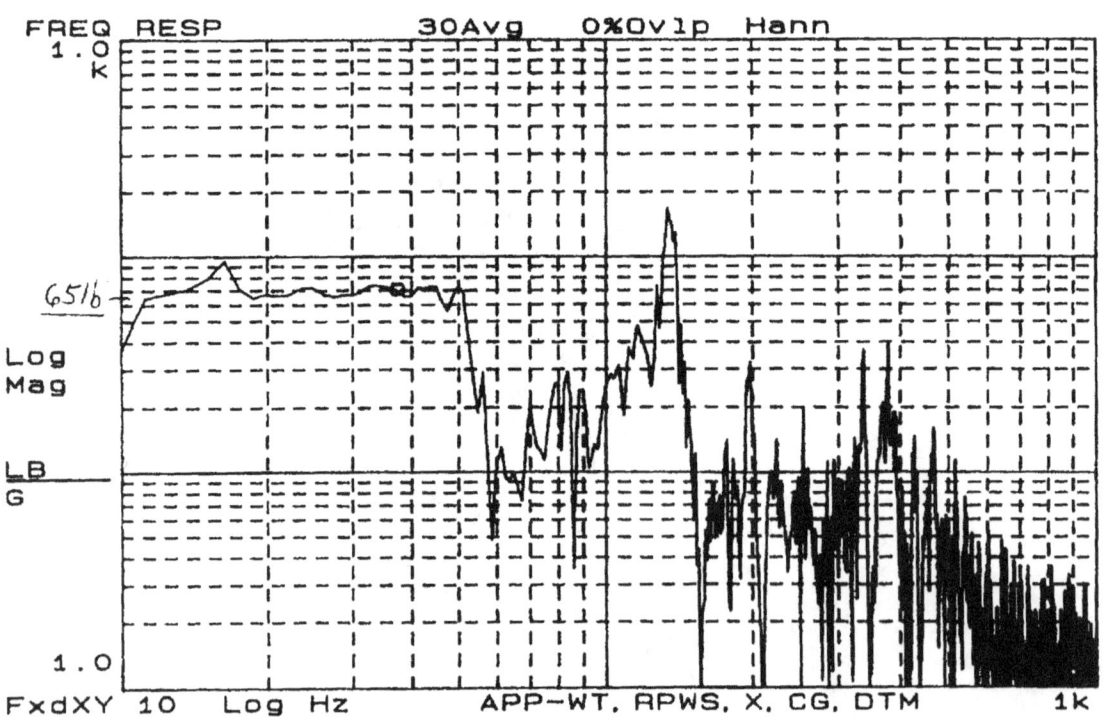

FIGURE 25. Ratio of Lateral External Force To Lateral Acceleration at Nominal CG for Cassini
RPWS Instrument in DTM Spacecraft Acoustic Test

FIGURE 26. CRIS Instrument from Advanced Composition Explorer (ACE) Spacecraft
Mounted with Flight Force Gages on Shaker for Vertical Vibration Test

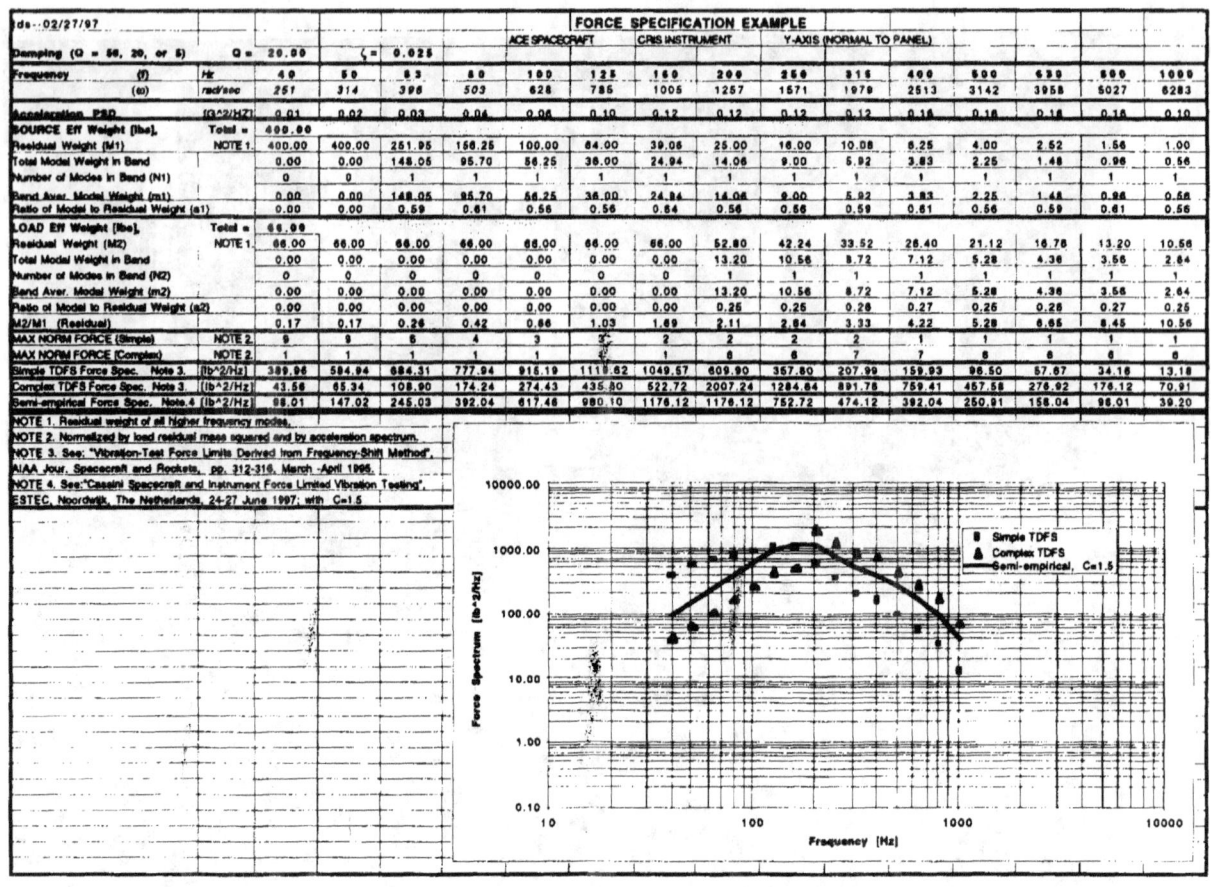

ds--02/27/97					**FORCE SPECIFICATION EXAMPLE**											
					ACE SPACECRAFT		CRIS INSTRUMENT		Y-AXIS (NORMAL TO PANEL)							
Damping (Q = 50, 20, or 5)	Q =	20.00	ζ =	0.025												
Frequency (f)	Hz	40	50	63	80	100	125	160	200	250	315	400	500	630	800	1000
(ω)	rad/sec	251	314	396	503	628	785	1005	1257	1571	1979	2513	3142	3958	5027	6283
Acceleration PSD	[G^2/HZ]	0.01	0.02	0.03	0.04	0.06	0.10	0.12	0.12	0.12	0.12	0.16	0.16	0.16	0.16	0.10
SOURCE Eff Weight [lbs],	Total =	400.00														
Residual Weight (M1)	NOTE 1	400.00	400.00	251.95	156.25	100.00	64.00	39.06	25.00	16.00	10.08	6.25	4.00	2.52	1.56	1.00
Total Modal Weight in Band (N1)		0.00	0.00	148.05	95.70	56.25	36.00	24.94	14.06	9.00	5.92	3.83	2.25	1.48	0.96	0.56
Number of Modes in Band (N1)		0	0	1	1	1	1	1	1	1	1	1	1	1	1	1
Band Aver. Modal Weight (m1)		0.00	0.00	148.05	95.70	56.25	36.00	24.94	14.06	9.00	5.92	3.83	2.25	1.48	0.96	0.56
Ratio of Modal to Residual Weight (a1)		0.00	0.00	0.59	0.61	0.56	0.56	0.64	0.56	0.56	0.59	0.61	0.56	0.59	0.61	0.56
LOAD Eff Weight [lbs],	Total =	66.00														
Residual Weight (M2)	NOTE 1	66.00	66.00	66.00	66.00	66.00	66.00	66.00	52.80	42.24	33.52	26.40	21.12	16.78	13.20	10.56
Total Modal Weight in Band (N2)		0.00	0.00	0.00	0.00	0.00	0.00	0.00	13.20	10.56	8.72	7.12	5.28	4.36	3.56	2.64
Number of Modes in Band (N2)		0	0	0	0	0	0	0	1	1	1	1	1	1	1	1
Band Aver. Modal Weight (m2)		0.00	0.00	0.00	0.00	0.00	0.00	0.00	13.20	10.56	8.72	7.12	5.28	4.36	3.56	2.64
Ratio of Modal to Residual Weight (a2)		0.00	0.00	0.00	0.00	0.00	0.00	0.00	0.25	0.25	0.26	0.27	0.25	0.26	0.27	0.25
M2/M1 (Residual)		0.17	0.17	0.26	0.42	0.66	1.03	1.69	2.11	2.64	3.33	4.22	5.28	6.65	8.45	10.56
MAX NORM FORCE (Simple)	NOTE 2	9	9	6	4	3	2	2	2	2	2	1	1	1	1	1
MAX NORM FORCE (Complex)	NOTE 2	1	1	1	1	1	1	6	6	6	7	7	8	8	6	6
Simple TDFS Force Spec. Note 3.	[lb^2/Hz]	389.96	584.94	684.31	777.94	916.19	1118.62	1049.57	609.90	357.80	207.99	159.93	96.50	57.67	34.16	13.18
Complex TDFS Force Spec. Note 3.	[lb^2/Hz]	43.56	65.34	108.90	174.24	274.43	435.80	522.72	2007.24	1284.64	891.78	759.41	457.56	276.92	176.12	70.91
Semi-empirical Force Spec. Note 4	[lb^2/Hz]	98.01	147.02	245.03	392.04	617.46	980.10	1176.12	1176.12	752.72	474.12	392.04	250.91	158.04	98.01	39.20

NOTE 1. Residual weight of all higher frequency modes.
NOTE 2. Normalized by load residual mass squared and by acceleration spectrum.
NOTE 3. See: "Vibration-Test Force Limits Derived from Frequency-Shift Method", AIAA Jour. Spacecraft and Rockets, pp. 312-316, March -April 1995.
NOTE 4. See:"Cassini Spacecraft and Instrument Force Limited Vibration Testing", ESTEC, Noordwijk, The Netherlands, 24-27 June 1997; with C=1.5

FIGURE 27. Example of Spread Sheet for Calculating Force Limits--ACE Spacecraft CRIS Instrument

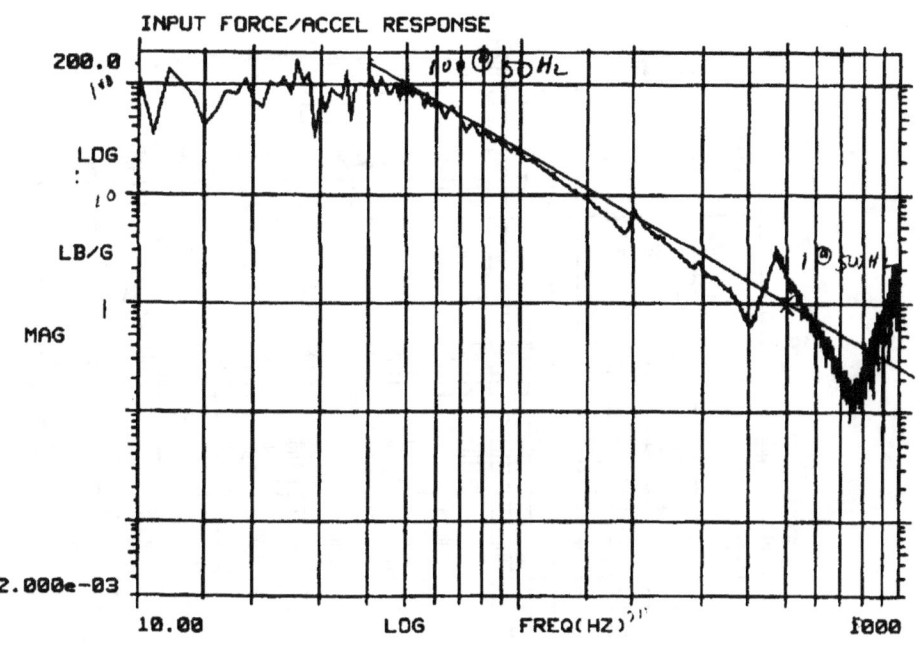

FIGURE 28. Tap Test Measurement of the Apparent Mass of the ACE Spacecraft Honeycomb Panel at One of the CRIS Instrument Mounting Positions

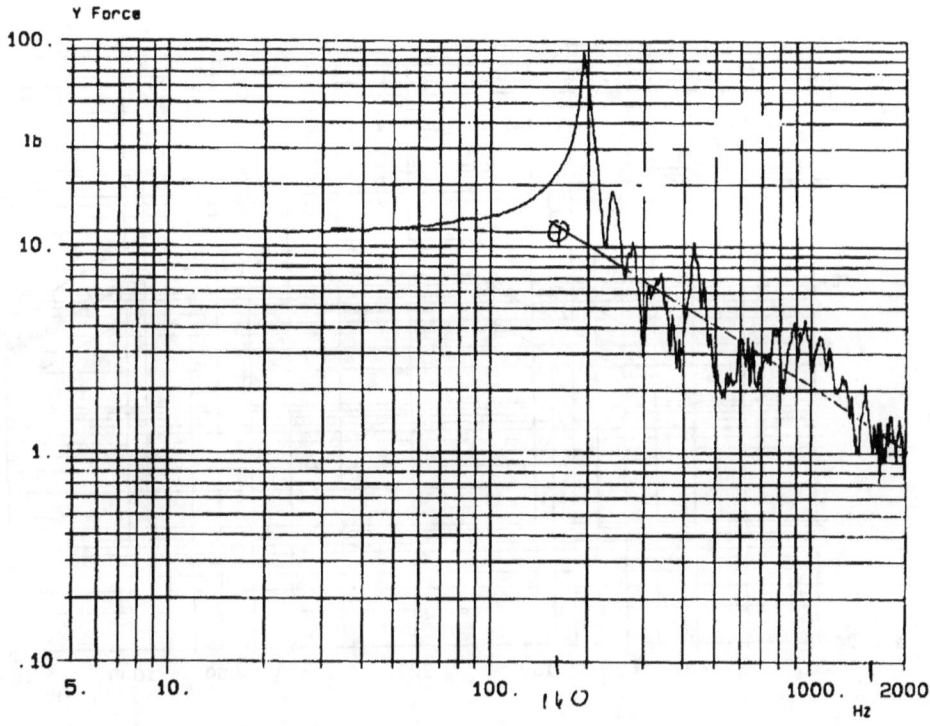

FIGURE 29. Shaker Measurement of the Apparent Mass of the CRIS Instrument in 0.25 G Sine Sweep Test (Multiply ordinate by four to obtain apparent mass.)

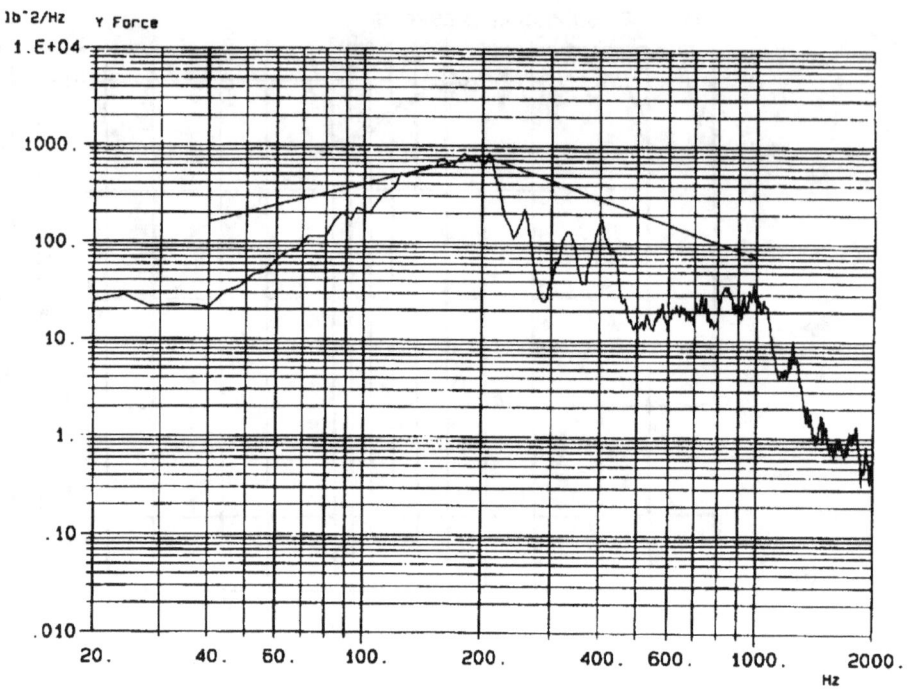

FIGURE 30. Total Vertical Force Limited to Specification in Protoflight Level Vertical Random Vibration Test of CRIS Instrument

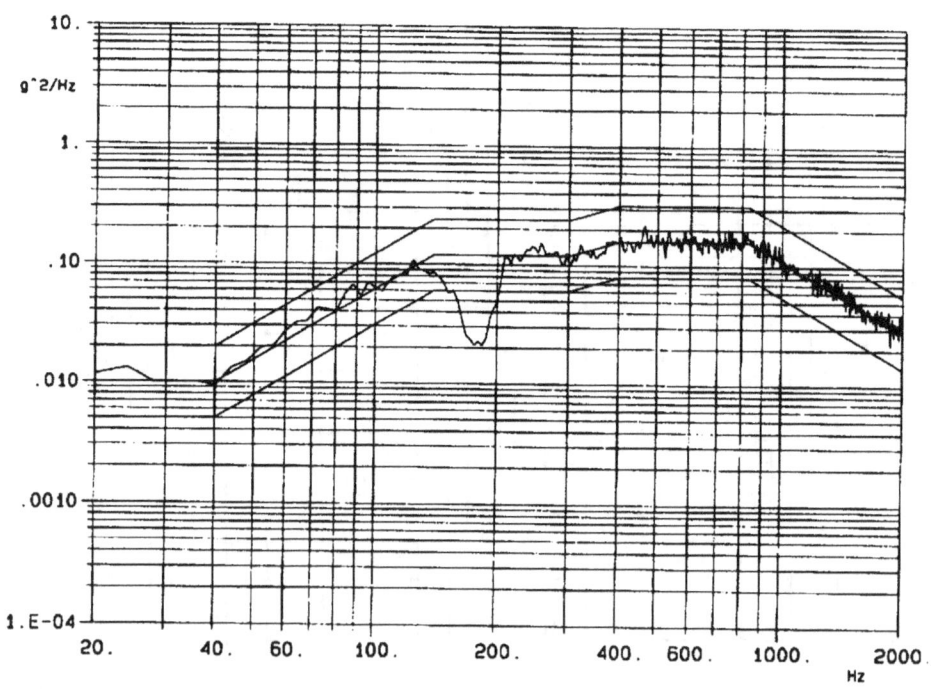

FIGURE 31. Notched Acceleration Input in Protoflight Level Vertical Random Vibration Test of CRIS Instrument

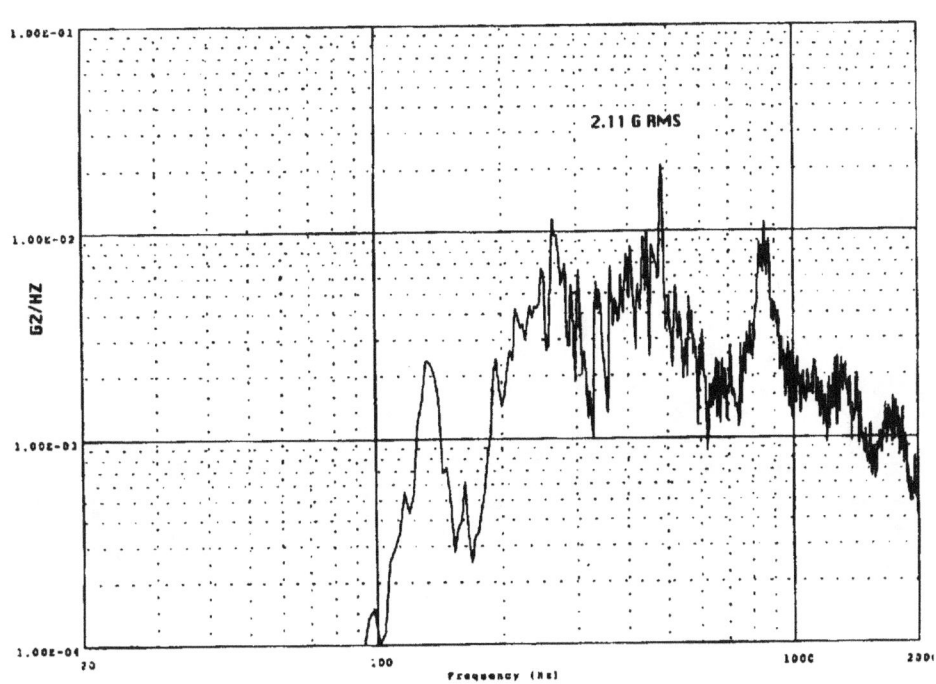

FIGURE 32. Acceleration Spectrum Measured Normal to Panel Near Mounting Foot of CRIS
Instrument in Protoflight Level Acoustic Test of ACE Spacecraft

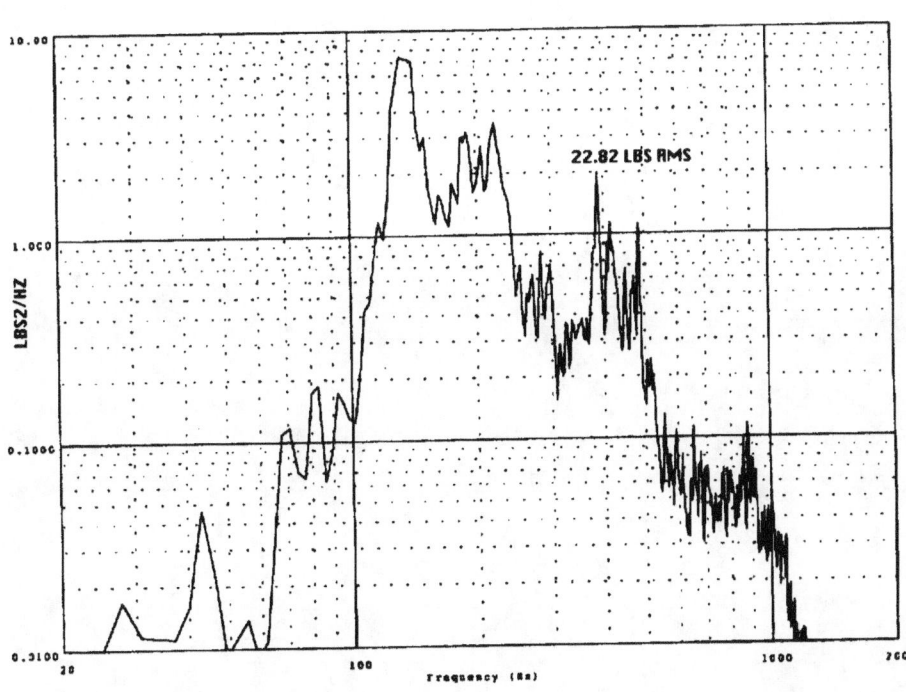

FIGURE 33. Total Normal Force Spectrum Measured Under Twelve Mounting Feet of CRIS
Instrument in Protoflight Level Acoustic Test of ACE Spacecraft

5.0 Instrumentation And Testing

Herein are described the characteristics and use of piezo-electric force transducers and other instrumentation employed in force limited vibration testing. Some important considerations in the planning and conduct of the tests are also discussed.

5.1 Piezo-electric Force Transducers

The use of piezo-electric force transducers for force limited vibration testing is highly recommended over other types of force measurement means such as strain transducers, armature current, weighted accelerometers, etc. The historical review in Section 2 teaches that the basic concepts and methods of force limited vibration testing were espoused and recommended, some twenty or thirty years ago. It is the authors belief that the primary reason that force limiting has not been previously accepted and implemented on a routine basis was the lack of a practical force measurement device, prior to the development of the piezo-electric, tri-axial force transducer. The advent of these transducers has made the measurement of force in vibration tests almost as convenient and accurate as the measurement of acceleration.

The high degree of linearity, dynamic range, rigidity, and stability of quartz make it an excellent piezo-electric material for both accelerometers and force transducers [16]. Similar signal processing, charge amplifiers and voltage amplifiers, may be used for piezo-electric force transducers and accelerometers. However, there are several important differences between these two types of measurement. Force transducers must be inserted between (in series with) the test item and shaker and therefore they require special fixtures, whereas accelerometers are placed upon (in parallel with) the test item or shaker. The total force into the test item from several transducers placed at each shaker attachment may be obtained by simply using a junction to add the charges before they are converted to voltage. On the other hand, the output of several accelerometers is typically averaged rather than summed. Finally, piezo-electric force transducers tend to put out more charge than piezo-electric accelerometers because the force transducer crystals experience higher loading forces, so sometimes it is necessary to use a charge attenuator between the force transducer and the charge amplifier.

5.1.1 Force Transducer Preload

Piezo-electric force transducers must be preloaded so that the transducer always operates in compression. The transverse forces are carried through the force transducer by friction forces. These transverse forces act internally between the quartz disks inside the transducer as well as between the exterior steel disks and the mating surfaces. Typically the maximum transverse load is 0.2, the coefficient of friction, times the compressive preload. Having a high preload, and smooth transducer and mating surfaces, also minimizes several common types of transducer measurement errors, e.g. bending moments being falsely sensed as tension/compression if gapping occurs at the edges of the transducer faces. However, using flight hardware and fasteners, it is usually impossible to achieve the manufacturers recommended preload, so some calculations are necessary to insure proper performance. Sometimes it is necessary to trade-off transducer capability for preload and dynamic load. (This is often the case if there are large dynamic moments which can't be eliminated by designing the fixtures to align the load paths.) The three requirements for selecting the preload are: 1. it must be sufficient to carry the transverse loads through the transducer by friction, 2. it must be sufficient to prevent loss of compressive preload at any point on the transducer faces due to the dynamic forces and moments, and 3. it must be limited so that the maximum stress on the transducer does not exceed that associated with the manufacturer's recommended maximum load configuration.

Transducer preloading is applied using a threaded bolt or stud which passes through the inside diameter of the transducer. With this installation, the bolt or stud acts to shunt past the transducer a small portion of any subsequently applied load, thereby effectively reducing the transducer's sensitivity. Calibration data for the installed transducers is available from the manufacturer if they are installed with the manufacturer's standard mounting hardware. Otherwise, the transducers must be calibrated in situ as discussed in the next section.

5.1.2 Force Transducer Calibration

The force transducer manufacturer provides a nominal calibration for each transducer, but the sensitivity of installed units depends on the size and installation of the bolt used for preloading and therefore must be calculated or measured in situ. This may be accomplished either quasi-statically or dynamically. Using the transducer manufacturer's charge amplifiers and a low noise cable, the transducers will hold their charge for many hours, so that it is possible to calibrate them statically with weights or with a hydraulic loading machine. If weights are used, it is recommended that the calibration be performed by loading the transducers, re-setting to short-out the charge, and then removing the load, in order to minimize the transient overshoot.

The simplest method of calibrating the transducers for a force limited vibration test is to conduct a preliminary low-level sine sweep or random run and to compare the apparent mass measured at low frequencies with the total mass of the test item. The appropriate apparent mass is the ratio of total force in the shaker direction to the input acceleration. The comparison must be made at frequencies much lower than the first resonance frequency of the test item. Typically the measured force will be approximately 80 to 90% of the weight in the axial direction and 90 to 95% of the weight in the lateral directions, where the preloading bolts are in bending rather than in tension or compression. Alternately, the calibration correction factor due to the transducer preloading bolt load path may be calculated by partitioning the load through the two parallel load paths according to their stiffness; the transducer stiffness is provided by the manufacturer, and the preload bolt stiffness in tension and compression or bending must be calculated. (The compliance of any structure in the load path between the bolt and transducer must be added to the transducer compliance.)

5.1.3 Force Transducer Signal Conditioning

It is strongly recommended that the <u>total</u> force in the shaker excitation direction be measured in a force limited vibration test. The total force from a number of transducers in parallel is readily obtained using a junction box which effectively sums the charges, and therefore the forces, before conditioning the signal with a charge amplifier. (An alternative is to specify limits for the force at individual attachment positions as in the case history in Section 6.1.2.) The same charge amplifiers used for piezo-electric accelerometers may be used for force transducers. However, the charge amplifiers made expressly for force transducers offer a choice of time constants, so that quasi-static (time constants of many hours) measurements of bolt preload, etc. may be obtained, as well as the dynamic measurements one usually associates with piezo-electric transducers. Also the charge amplifiers made expressly for force transducers usually have the capability to accommodate higher charge inputs, which are characteristic of force transducers. However, charge attenuators are readily available if they are needed.

Since vibration tests are normally conducted sequentially in three perpendicular axes, it is convenient to employ tri-axial force transducers. In addition, it is sometimes necessary to limit the cross-axis force and the moments in addition to the in-axis force; this is particularly the case in tests of large eccentric test items such as spacecraft. For these applications, the six force resultant forces and moments for a single node may be measured with a combination, commonly four, of tri-axial force transducers and commercially available voltage summers manufactured expressly for this purpose.

5.2 Test Fixtures

5.2.1 Design Concepts

The preferred method of configuring the force transducers is to sandwich one transducer between the test item and conventional test fixture at each attachment position and use fasteners which are longer than the conventional ones to accommodate the height of the transducers. In this configuration, there is no fixture weight above the transducers and the transducer force is identical to the force into the test item. Sometimes the preferred approach is impractical, e.g. if there are too many attachment points or the attachments involve shear pins in addition to bolts. In these cases it may be necessary to use one or more light-weight intermediate adapter plates as an interface between the test item and the force transducers. For example, if the test item mounts at three feet and each foot involves two bolts and a shear pin, a candidate design would be to have a small plate attached to a big stud for each foot. The small plate would pick up the two mounting bolts and shear pin, and the stud would go through a medium sized force transducer into a shaker adapter plate. Alternately, if the mounting configuration involves sixteen small bolts in a circular pattern, the fixture might consist of one intermediate ring which accepts the sixteen small bolts and is mounted on eight equally spaced force transducers

5.2.2 Fixture Weight Guideline

The recommendation is that the total weight of any intermediate adapter plates above the force transducers do not exceed ten percent of the weight of the test item. This limitation is necessary because the force transducers read the sum of the force required to accelerate the interface plate and that delivered to the test item. If the fixture weight exceeds the 10% criterion, force limiting will only be useful for the first one or two modes in each axis. Use of a circuit to subtract the interface plate force in real time, is not recommended because of the errors that result when the interface plate is not rigid. The use of armature current to measure shaker force is also not generally useful, because the weight of the armature and fixture typically are much greater than 10% of that of the test item.

5.2.3 Mass Cancellation

If an intermediate plate is used and the plate moves unilaterally as a rigid body, its acceleration may be measured with an accelerometer and subtracted, in real time from the gross force measured by the transducers underneath the intermediate plate to obtain the net force delivered to the test item. This approach has been utilized in a number of relatively simple tests, but is not recommended because of the possibility and consequences of errors. As the frequency approaches a resonance frequency of the intermediate plate on its mounting, the phase angle of the plate acceleration and applied force changes such that the aforementioned cancellation scheme does not work and may make matters worse. Also, the additional instrumentation needed to implement the mass cancellation scheme gives rise to increased electrical noise and possibility of set-up error.

5.3 Testing Considerations

5.3.1 Criteria For Force Limiting

The purpose of force limiting is to reduce the response of the test item at its resonances on the shaker in order to replicate the response at the combined system resonances in the flight mounting configuration. Force limiting is most useful for structure-like test items which exhibit distinct, lightly damped resonances on the shaker. Examples are complete spacecraft, cantilevered structures like telescopes and antennas, lightly damped assemblies such as cold stages, fragile optical components, and equipment with pronounced fundamental modes such as a rigid structure with flexible feet. The amount of relief available from force limiting is greatest when the structural impedance (effective mass) of the test item is equal to, or greater than, that of the mounting structure. However, it is recommended that notches deeper than 14 dB not be implemented without appropriate peer review. Force limiting is most beneficial when the penalties of an artificial test failure are high. Sometimes this is after an initial test failure in a screening type of test.

5.3.2 Test Planning

Several considerations need to be addressed in the test planning if force limiting is to be employed. First the size, number, and availability of the force transducers need to be identified as well as any special fixture requirements to accommodate the transducers. Next, the approach for deriving and updating the force specification needs to be decided. Finally the control strategy must be decided and written into the test plan. Special cases may include cross-axis force, moment, individual force, and response limiting in addition to or in lieu of the in-axis force. In some instances, the control strategy will be limited by the control system capabilities. In all cases, it is recommended that the control strategy be kept as simple as possible, in order to expedite the test and to minimize the possibility of mistakes.

Accelerometers on the fixture are also required in force limited vibration tests in order to control the acceleration input to the acceleration specification at frequencies other than at the test item resonances. In addition, it is often convenient to use a limited number of accelerometers to measure the response at critical positions on the test item. These response accelerometers may be used only for monitoring or, if justified by appropriate rationale, for response limiting in addition to the force limiting.

5.3.3 Cost And Schedule

Project managers often inquire regarding the cost and schedule impact of doing force limiting. Once force limiting has become routine in the vibration testing laboratory, its use definitely results in both cost and schedule savings. The primary savings is through the prevention of unnecessary, overtest failures, which adversely impact both cost and schedule. Implementation of force limiting on a routine basis also eliminates most of the contentious discussions about test levels. Also the effort previously required to measure, analyze, and limit numerous responses in complex vibration tests is greatly reduced by force limiting, as illustrated by the Cassini spacecraft test example in Section 6.2.

The first or second time that an organization employs force limiting, there will naturally be some down time and slow going. The first consideration must be the availability of force transducers of the appropriate size. Tri-axial force transducers are relatively expensive, compared to accelerometers, and are sometimes long-lead procurement items, so an assortment of transducers are usually procured over time by several projects and maintained by the vibration test laboratory for use on future projects. If the first application is a flight project, it is recommended that some help be sought from the sponsor or another organization that has experience in force limiting. The only other potential additional cost of using force limiting is the development of special fixtures. Usually the force transducers can be utilized simply by placing one at each mounting position and using a longer bolt to accommodate the thickness of the transducer. Configurations where special fixtures may be needed are: those which involve a large number of mounting points, say more than twelve, or those with shear pins or complicated fittings, such as the flight latches in the example of Section 6.1.2.

5.3.4 Specification of Force Limits

Force limits are analogous and complementary to the acceleration specifications used in conventional vibration testing. Just as the acceleration specification is the frequency spectrum envelope of the in-flight acceleration at the interface between the test item and flight mounting structure, the force limit is the envelope of the in-flight force at the interface. In force limited vibration tests, both the acceleration and force specifications are needed, and the force specification is proportional to the acceleration specification. Therefore force limiting does not compensate for errors in the development of the acceleration specification, e.g. undue conservatism or lack thereof. These errors will carry over into the force specification. Since in-flight vibratory force data are lacking, force limits are usually derived from coupled system analyses and impedance information obtained from measurements or finite element models (FEM). Also, considerable data on the interface force between spacecraft and components are becoming available from spacecraft acoustic tests, and semi-empirical methods of predicting force limits are available.

Force spectra have typically been developed in one-third octave bands (see example in Section 4.4), but other bandwidths, e.g. octave or one-tenth octave bands, may also be used. Force limiting may usually be restricted to an upper frequency encompassing approximately the first three modes in each axis; which might be approximately 100 Hz for a large spacecraft, 500 Hz for an instrument, or 2000 Hz for a small component. It is important to take into account that the test item resonances on the shaker occur at considerably higher frequencies than in flight. Therefore care must be taken not to roll off the force specification at a frequency lower than the fundamental resonance on the shaker and not to roll off the specification too steeply, i.e. it is recommended that the roll-offs of the force spectrum be limited to approximately 9 dB/octave.

5.3.5 Vibration Controllers

Most of the current generation of vibration test controllers have the two capabilities needed to implement force limiting. First, the controller must be capable of extremal control, sometimes called maximum or peak control by different vendors. In extremal control, the largest of a set of signals is limited to the reference spectrum. (This is in contrast to the average control mode in which the average of a set of signals is compared to the reference signal.) Most controllers used in aerospace testing laboratories support the extremal control mode. The second capability required is that the controller must support different reference spectra for the response limiting channels, so that the force signals may have limit criteria specified as a function of frequency. Controllers which support different reference spectra for limit channels are now available from most venders and in addition upgrade packages are available to retrofit some of the older controllers for this capability. If the controller does not have these capabilities, notching of the acceleration specification to limit the measured force to the force specification must be done manually in low level runs.

5.3.6 Low-Level Runs

It is advantageous to keep the number of test runs as low as feasible, both to save testing time and to avoid accumulating unnecessary fatigue of the test hardware. A low-level sine-sweep or random run with a flat frequency spectrum is often conducted before and after the high-level vibration run in each axis to measure the vibration signature for "health" monitoring. The reaction force is an excellent choice for the health monitoring signature, and the ratio of force to input acceleration measured in the "before" run can also be used to update the effective masses and resonance frequencies used to derive the force specification. (Sometimes the derivation of the force specification is deferred until this data become available.)

It is also advantageous to conduct two low-level runs (often -18 dB) with the same input acceleration spectral shape as the high-level run; the first without force limiting and the second with force limiting, using a scaled down force limit. Comparison of the forces measured in these two low-level runs with the amount of notching achieved in the second run with force limiting provides verification that the force limit is appropriate and that the notching is as it should be. Extrapolation of measured responses to full level and comparison with the loads criteria and any adjustments to the force specification should take place after these two runs. After it is determined that the results are satisfactory, any intermediate-level runs and the full-level run may be conducted. With the new controllers, it is becoming common practice to come "on the air" at a low level and then proceed through the intermediate levels to full level, without shutting down. Thus ideally, each axis may be conducted with no more than five runs: 1) low-level pre-test signature, 2) -18 dB without force limiting, 3) -18 dB with force limiting, 4) "on the air" at intermediate level and progressing to full level, and 5) low-level post-test signature.

6.0 Case Histories

The three random vibration tests selected as case histories include a component, an instrument, and a spacecraft. These test items span a mass range greater than 10^5. The component test was conducted in March of 1991 and the spacecraft test in November of 1996; the technology, particularly as regards the prediction of force limits and the shaker control, progressed considerably in this time interval.

6.1 Hubble Wide Field Planetary Camera II AFM Component and Complete Instrument

6.1.1 Articulating Fold Mirror Component Random Vibration Test [42]

There are three articulating fold mirror (AFM) assemblies in the Wide Field Planetary Camera II (WFPCII) installed in the Hubble Telescope during the first servicing mission in December 1993. The role of the AFM is to provide a means for very accurate on-orbit alignment of the optical beam on the secondary relay mirrors which contain the correction for the Hubble primary mirror spherical aberration. The photograph in Fig. 34 shows an AFM before the mirror is coated. The AFM utilizes three small electro-strictive actuators to articulate the mirror. By necessity, the AFM is small and delicate; the total unit weighs approximately 100 gm and the articulating portion, the mirror and bezel, weighs less than 30 gm.

Figure 35 shows an AFM mounted in the vibration test fixture designed to accommodate three small (approx. 2.5 cm square) commercially available tri-axial force transducers. Conical spacers are used to mate the transducers with the #2 screws which normally attach the AFM to a bulkhead of the optical bench of the WFPCII. The apparent mass normal to the WFPCII bulkhead was measured with an impact hammer to be 3.2 kg at 800 Hz, the first axial resonance of the AFM. The force specification was determined using the simple TDFS method. Notice from Fig. 5 that even with the small ratio (0.01) of AFM mass to bulkhead effective mass and with the measured Q of 50, the normalized ratio of force spectral density to acceleration spectral density of (100) is still 14 dB less than Q squared (2500), which is the amplification expected in a conventional vibration test.

The results of the Z-axis protoflight random vibration test of the AFM qualification unit are shown in Fig. 36. The unnotched input acceleration spectral value of 0.004 G²/Hz at 800 Hz was determined from measurements during the acoustic test of the optical bench of WFPCI, the original camera on the Hubble Space Telescope. Figure 36 also shows the total axial force input to the AFM, i.e. the sum of the Z axis outputs of the three force transducers in Fig. 35. The force shown in Fig. 36 is actually scaled up from a low level (18 dB down) test without force limiting to show what the force would have been in the full level test if the force were not limited. In the signal conditioning, the force feedback signal was multiplied by the ratio of the acceleration specification to the force specification so that extremal control of the both force and acceleration could be implemented by comparing both signals to the acceleration specification, as discussed in Section 5.3.2. For this reason, the force signal in Fig. 36 can be compared directly with the acceleration specification. Figure 36 shows that in the protoflight-level test with force limiting, the controller automatically notched the acceleration input by the amount the unlimited force signal would have exceeded its specification, i.e. about 10 dB, which is 4 dB less than that estimated from Fig. 5. The notch is very sharp and approximately the mirror image of the force peak. It is impractical to manually put in such sharp notches. Also without the force sensors to detect the frequency of the force peak, it would be difficult to place the notch at the correct frequency.

6.1.2 Wide-Field Planetary Camera II Instrument Random Vibration Test [42]

The complete WFPCII (weight 284 kg) was subjected to a vertical axis protoflight level random vibration test as shown in Fig. 37. A large (approx. 10 cm diam.) commercially available, tri-axial force transducer was located just below each of the three latches, which fasten the camera to the Hubble telescope. Force specifications for all three directions at all three latches were derived using the simple TDFS method with apparent mass data for the WFPCII and for the honeycomb container used to transport the WFPCII in the Space Shuttle to rendezvous with the orbiting telescope. The apparent masses of the container were measured by NASA GSFC with a modal impact hammer. The data for the vertical direction at the A latch of the container is shown in Fig. 38. (The A latch is on the far left in Fig. 37.) The three components (X, Y & Z) of force at each of the three latches were recorded and analyzed in a low level sine test of the WFPCII preceding the random vibration test, and the apparent mass of the WFPCII in the vertical direction at the A latch is shown in Fig. 39. After a low-level random run, it was determined that three of the four control channels available on the older controller were essential to control the high frequency acceleration input at each of the three latches, so only one control channel was available for force limiting. (New controllers have 32 to 64 control channels.) On the basis of analysis and the low level data, it was decided to limit the vertical force at the A latch which reacts most of the load because of the outboard (to the right in Fig. 37) center-of-gravity of the camera.

The acceleration measured at each of the three latches is compared with the acceleration specification for the protoflight random vibration test in Fig. 40, and the vertical force measured at each of the three latches is compared to the force specification for the A latch in Fig. 41. Comparison of Figs. 40 and 41 shows that limiting the A latch vertical force resulted in notching of the acceleration input at 35 Hz, 100 Hz and 150 Hz. The acceleration notch at 70 Hz was effected by reducing the vertical force specification at 70 Hz to compensate for the A latch transverse force not being in the control loop. Above 300 Hz, one of the three control accelerometers is equal to the specification at every frequency. Comparison of the notched vibration input in the AFM component vibration test described in the last section with the corresponding vibration levels measured a year later on the optical bench in the WFPCII system acoustic test confirmed that the vibration levels in the component test enveloped those in the system acoustic test, as they should for a valid component screening test.

6.2 Cassini Spacecraft System Random Vibration Test [28, 43]

Figure 42 is a photograph of the Cassini flight spacecraft mounted on the shaker for the vertical random vibration test which was conducted in November of 1997 [28, 43]. The Cassini orbiter weighs a total of 2,150 kilograms (4,750 pounds); after attaching the 350-kilogram Huygens probe (on the right in Fig. 42) and a launch vehicle adapter and loading more than 3,000 kilograms (6,600 pounds) of propellants, the spacecraft weight at launch is about 5,800 kilograms (12,800 pounds). The weight for the vibration test was somewhat less 3,809 kg (8,380 lb.), because the tanks were loaded to 60% of capacity with referee fluids. Because of the very dim sunlight at Saturn's orbit, solar arrays are not feasible and power will be supplied by a set of three Radioisotope Thermoelectric Generators (RTG's) which use heat from the natural decay of Most of plutonium to generate electricity to run Cassini. (Two of the RTG's are visible in the lower center and left of Fig. 42.) Twelve science experiments are carried onboard the Cassini orbiter and another six fly on the Huygens Titan probe. The schematic in Fig. 7 indicates a number of the

Cassini spacecraft instruments. the orbiter instruments are mounted on the Remote Sensing Platform at the upper left and on the Fields and Particles Platform at the upper right in Fig. 42.

Figure 43 shows the plan view of the spacecraft mounting ring before the spacecraft is attached to the shaker. The black offset weight positioned in the upper right quadrant of the ring is being subjected to vibration for moment proof testing of the shaker and mounting configuration. The spacecraft bolts to the ring at eight positions corresponding to the mounting feet locations on the spacecraft/ launch vehicle adapter. A large tri-axial force transducer is located under the mounting ring at each of these eight positions. The shaker fixture is restrained from moving laterally during the spacecraft vertical vibration test by three hydraulic bearings. The force capability of the shaker is about 35,000 lb. and virtually all of this capability was used to vibrate the spacecraft and shaker fixtures, which weighed about 6,000 lb. in addition to the 8380 lb. spacecraft.

Figure 44 compares the vibration test acceleration input specification with launch vehicle specifications and with data from a previous Titan launch vehicle flight. The acceleration specification was originally somewhat higher (0.04 G^2/Hz compared to 0.01 G^2/Hz). The specification was lowered in the 10 to 100 Hz frequency regime after reviewing the results of an extensive FEM pre-test analysis, which indicated that excessive notching would be required with the higher-level input. The specification was subsequently lowered in the 100 to 200 Hz regime as well, in order to accommodate the force capability of the shaker power amplifier, which was over ten years old and exhibited some instability problems during the two month period preceding the Cassini spacecraft vibration test. The resulting 0.01 G^2/Hz specification is less than the Booster Powered Phase specification at frequencies greater than approximately 80 Hz, but exceeds the Maximum Envelope of the TIV-07 Flight Data. The acceleration specification is defined in Fig. 45.

Figure 46 shows the force specification for the Cassini flight spacecraft random vibration test. The specification was derived by multiplying the acceleration specification in Fig. 45 by the squared weight of the spacecraft and by a factor of one-half. (This corresponds to the semi-empirical method discussed in Section 4.2.1 with a value of C = 0.707.) This value of C was selected on the basis of the pre-test analysis and in order to keep the proof test, which had a margin of 1.25 over the test limit loads, within the shaker force capability. The force specification was not rolled off at the shaker fundamental resonance as shown in Equation 29, because neither the pre-test analysis nor the actual vibration test data exhibited a distinct fundamental resonance of the spacecraft in the vertical axis. During the test, it was not necessary to modify or update the force specification specified in the test procedure.

Figures 47 and 48 respectively, show the input acceleration and force spectra measured in the actual full-level vibration test. Comparison of the measured acceleration spectra with the specification in Fig. 47 shows significant notching of ~8 dB at the probe resonance of approximately 17 Hz and of ~14 dB at the tank resonance of approximately 38 Hz. The responses at a number of critical positions on the spacecraft, as well as the other five components of the total input force vector, were monitored during the testing, but only the total vertical force signal was used in the controller feedback to notch the acceleration input. Comparison of the measured force with the specified force in Fig. 48 verifies that the force was at its limit at all the frequencies where notching occurred in the input acceleration.

Figures 49 and 50 show the acceleration inputs measured near the feet of a number of instruments mounted on the Fields and Particles and Remote Sensing Pallets, respectively. Comparison of these measured data with the random vibration test specifications for the instruments, which are also indicated in Figs. 49 and 50, demonstrates that many of the instruments reached their component vibration test levels in the spacecraft vibration test. The significant excedances below 50 Hz are covered by the instrument sine vibration test equipment. In addition, several major components of the spacecraft including the Huygens probe upper strut, the three RTG's, the magnetic canister struts, and the Fields and Particles Pallet struts reached their flight limit loads during the spacecraft vibration test. The only anomaly after the test, other than possibly those associated with spacecraft functional tests for which data are not available, was that the electrical resistance between the engineering model RTG and the spacecraft structure was measured after the test and found to be less than specified. The insulation between the RTG adapter bracket and the spacecraft was redesigned to correct this problem.

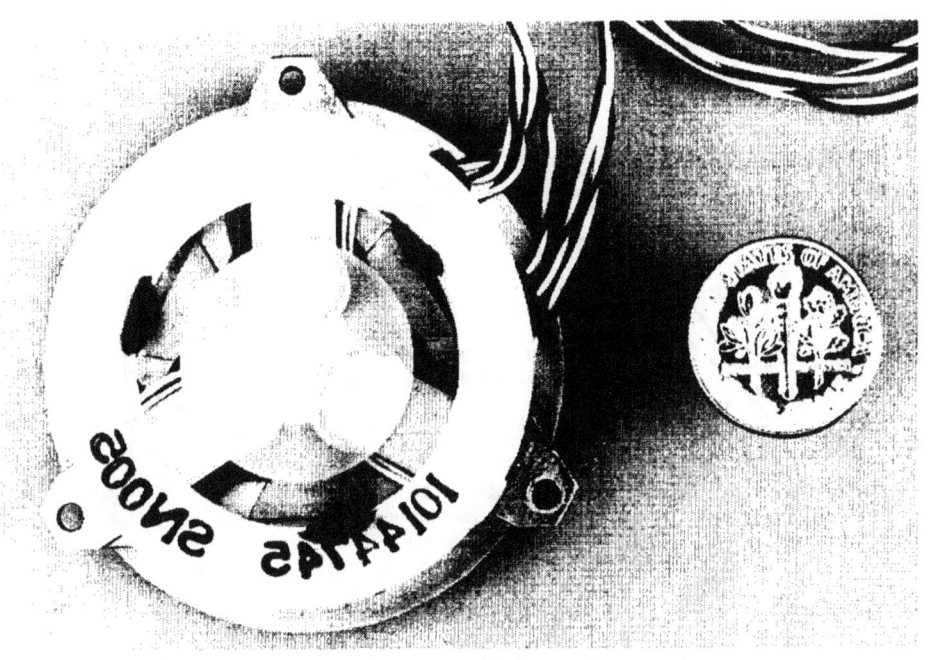

FIGURE 34. Photograph of 30-Gram Articulating Fold Mirror (AFM) Before Coating

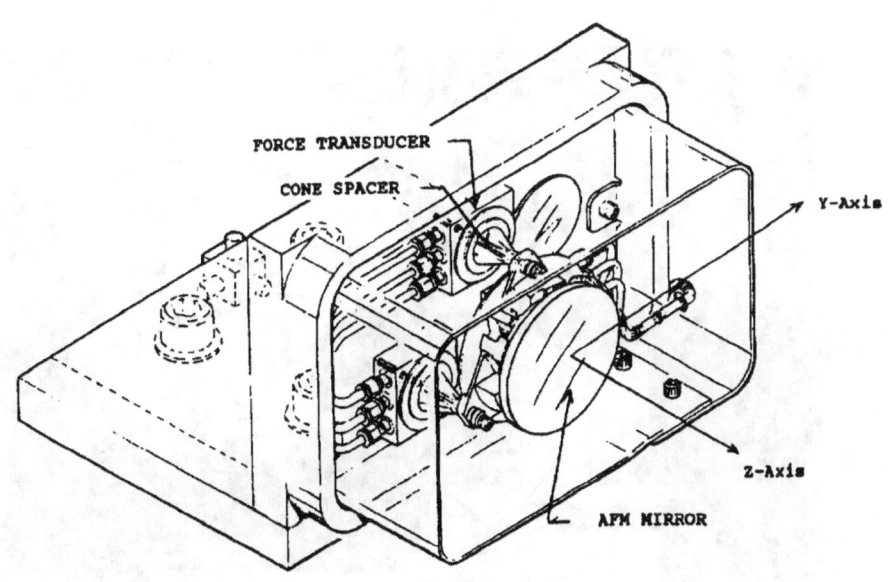

FIGURE 35. Sketch of Vibration Test Fixture for AFM Showing Mounting on Three Small
Tri-axial Force Gages Using Conical Adapters

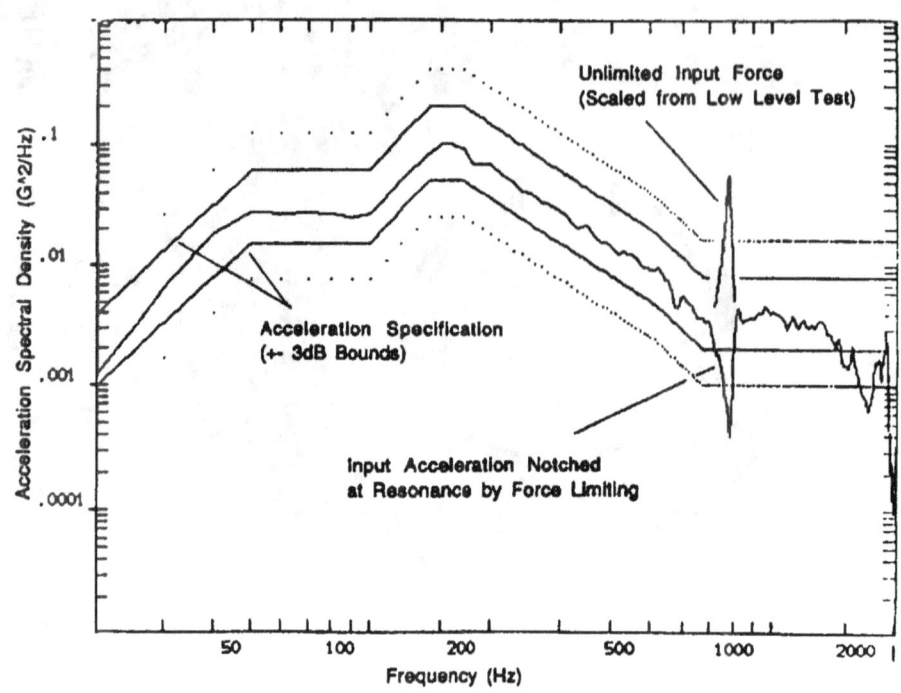

FIGURE 36. Notched Acceleration Input in Vertical Protoflight Random Vibration Test of the WFPCII Articulating Fold Mirror (AFM)

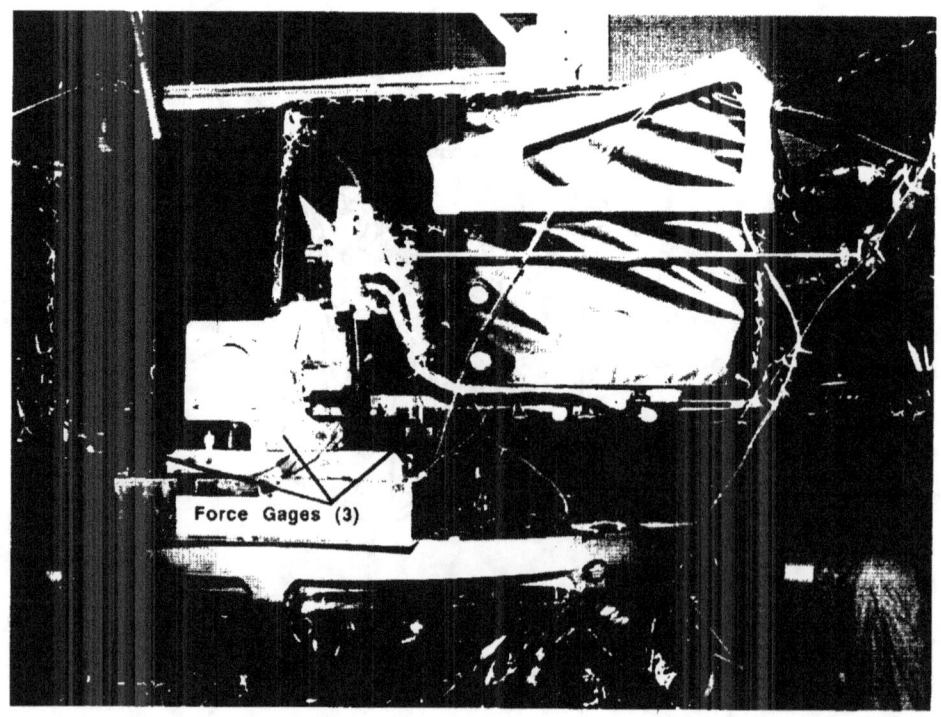

FIGURE 37. Hubble Space Telescope 600-Kilogram Wide Field Planatary Camera II (WFPCII) Mounted on Shaker for Vertical Axis Vibration Test

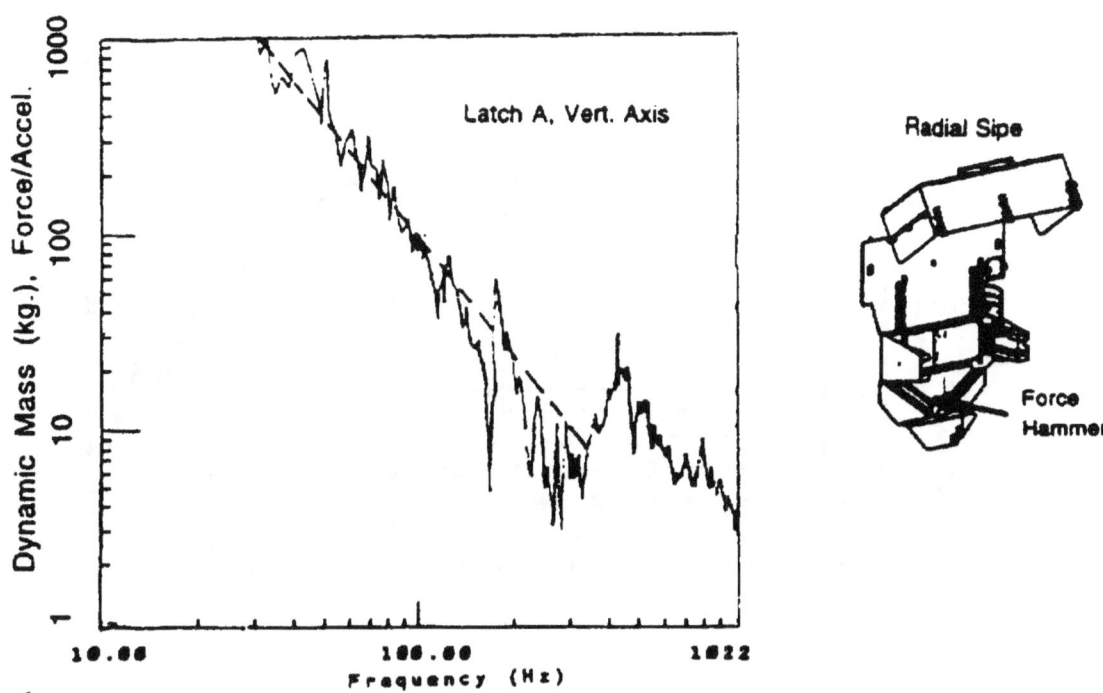

FIGURE 38. Apparant Mass of WFPCII Container at A Latch in Vertical Direction Measured with Modal Tap Hammer

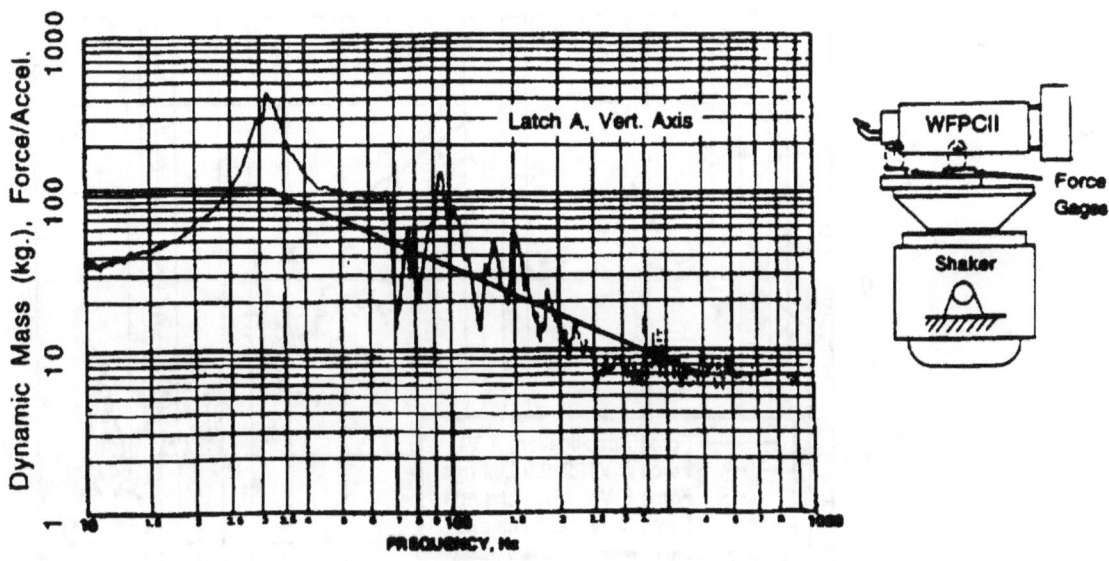

FIGURE 39. Apparant Mass of WFPCII at A Latch in Vertical Direction Measured on Shaker in Low-Level Sine-Sweep Vibration Test

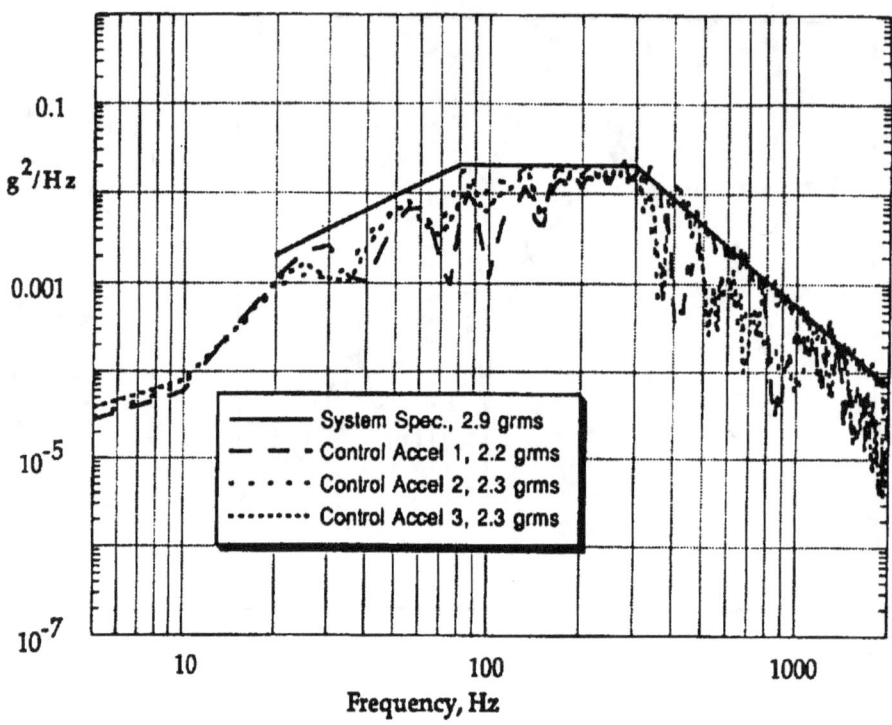

FIGURE 40. Measured and Specified Acceleration Input Spectra at Three Latches of the Wide Field Planetary Camera II in the Protoflight Random Vibration Test

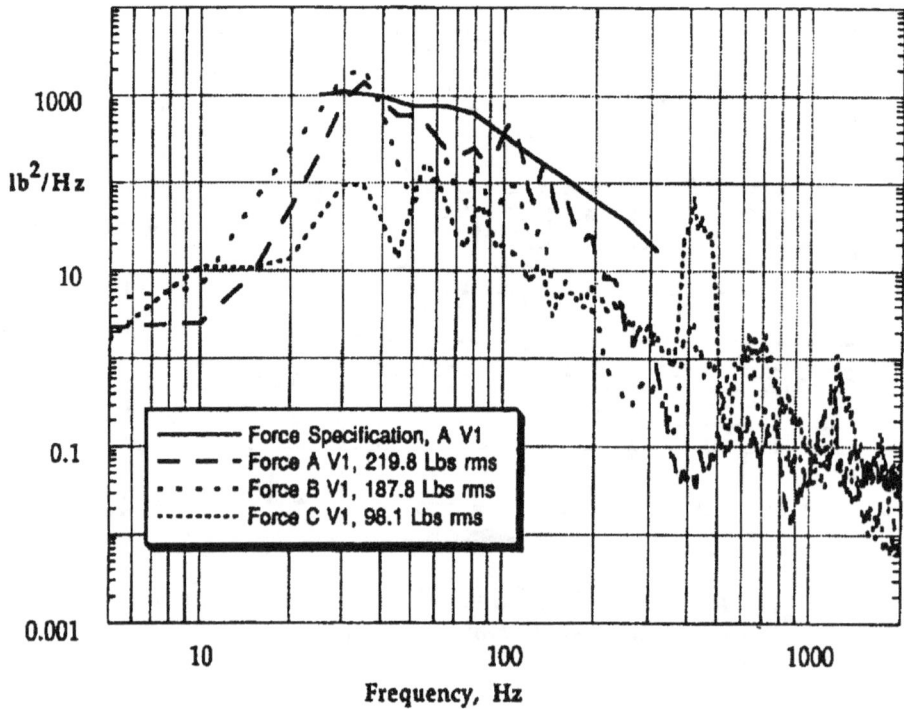

FIGURE 41. Measured and Specified Axial Force Spectra at Three Latches of the Wide Field Planetary Camera II in the Protoflight Random Vibration Test

FIGURE 42. Cassini Spacecraft Mounted on Shaker for Vertical Random Vibration Test

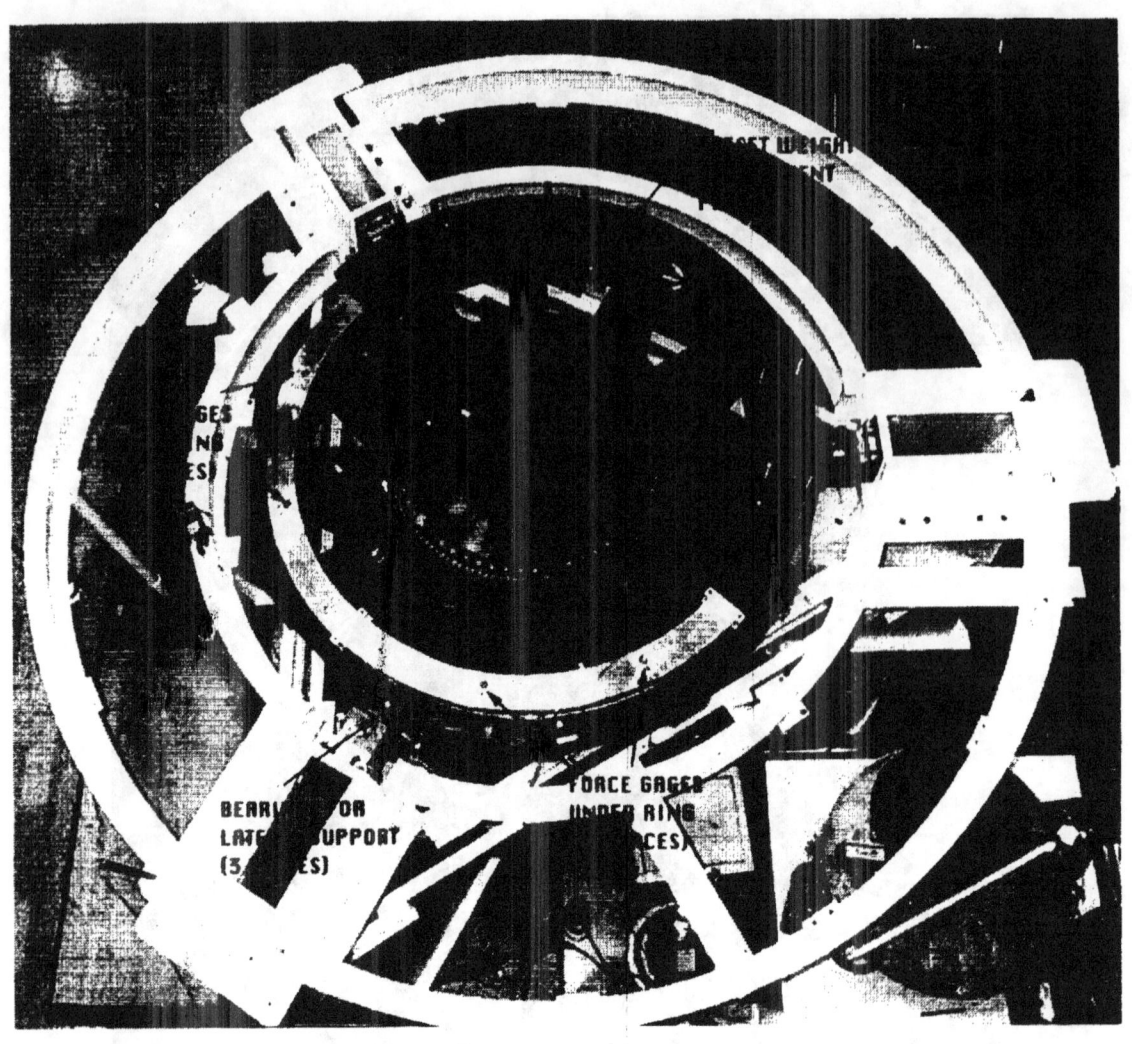

FIGURE 43. Plan View of Spacecraft Mounting Ring with Offset Weight for Moment Proof Test and of Lateral Restraint System for Cassini Spacecraft Vibration Test

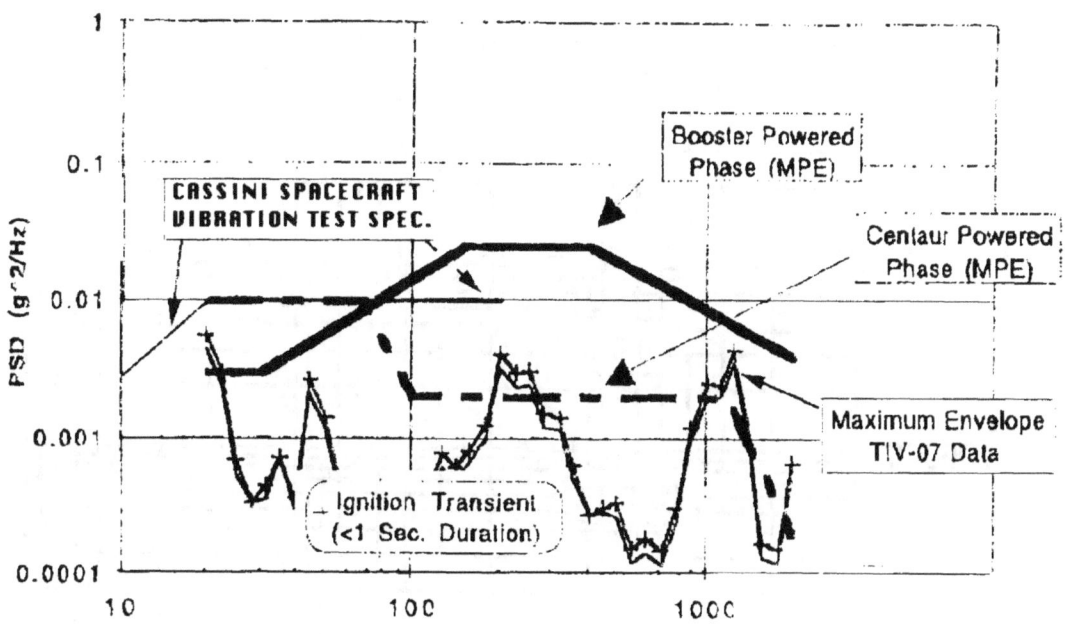

FIGURE 44. Comparison of Cassini Spacecraft Random Vibration Acceleration Input Specification with Launch Vehicle Specifications and Flight Data

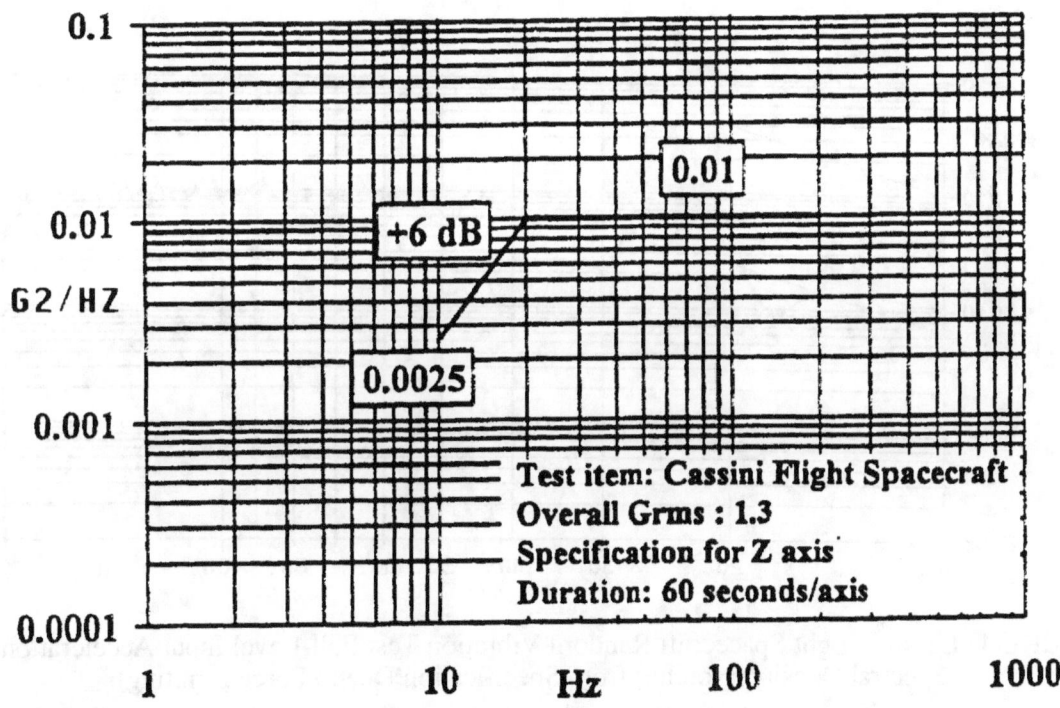

FIGURE 45. Cassini Flight Spacecraft Random Vibration Test Acceleration Specification

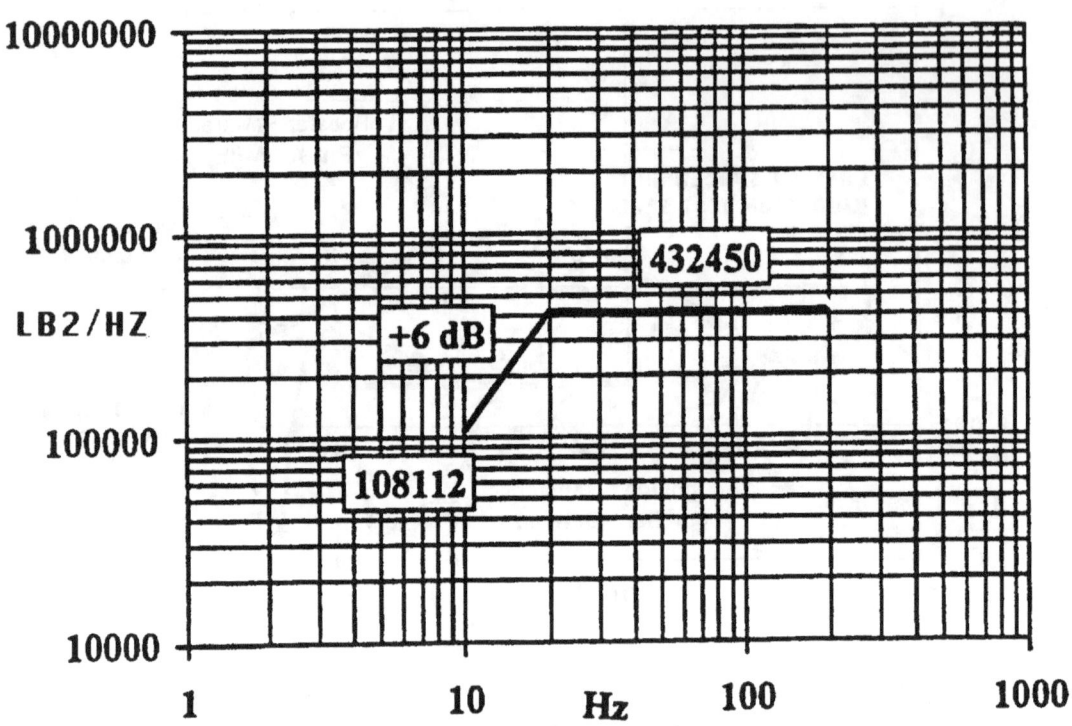

FIGURE 46. Cassini Flight Spacecraft Random Vibration Test Force Specification

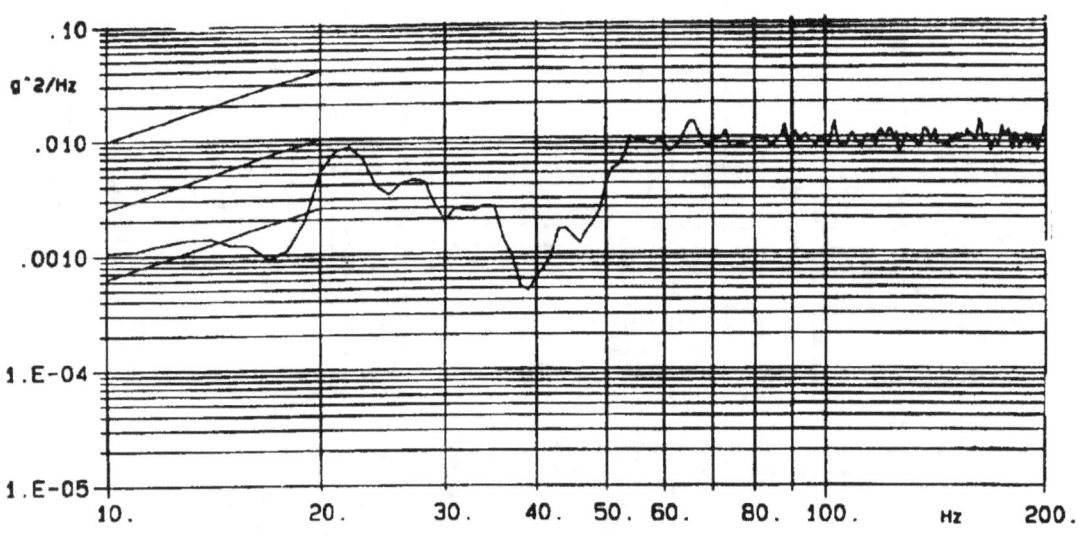

FIGURE 47 Cassini Flight Spacecraft Random Vibration Test Full-Level Input Acceleration
Spectral Density (Notches from Specification Due to Force Limiting)

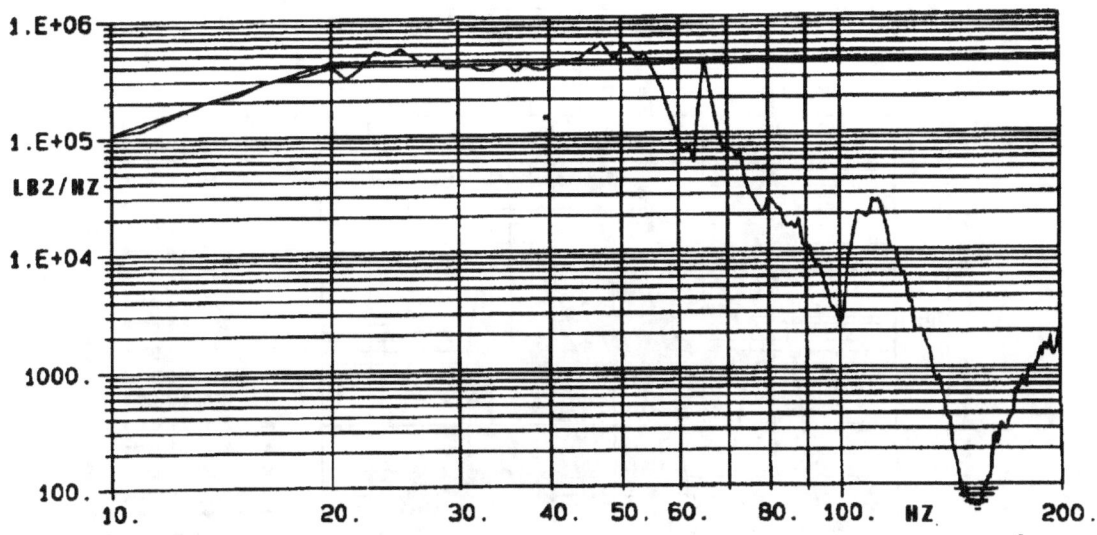

FIGURE 48. Total Vertical Force Measured in Cassini Flight Spacecraft Full-Level Random Vibration Test (Comparison with Vertical Force Limit Specification)

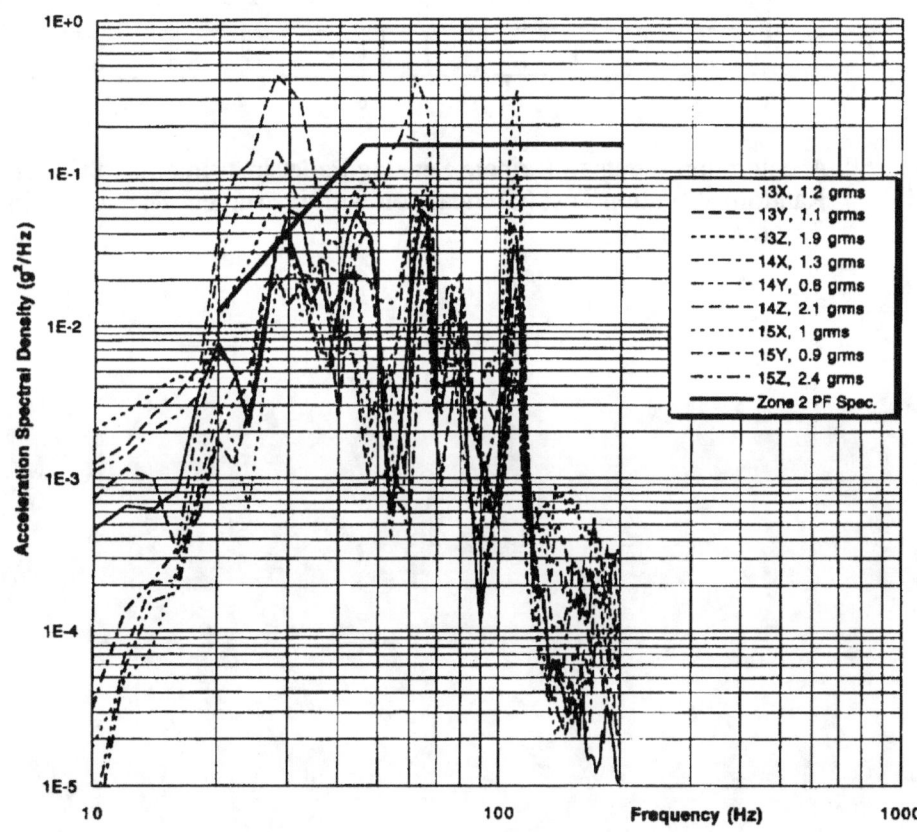

FIGURE 49. Acceleration Inputs to Fields and Particles Pallet Instruments in Cassini Full-Level Random Vibration Test (Comparison with Instrument Random Vibration Test Specification)

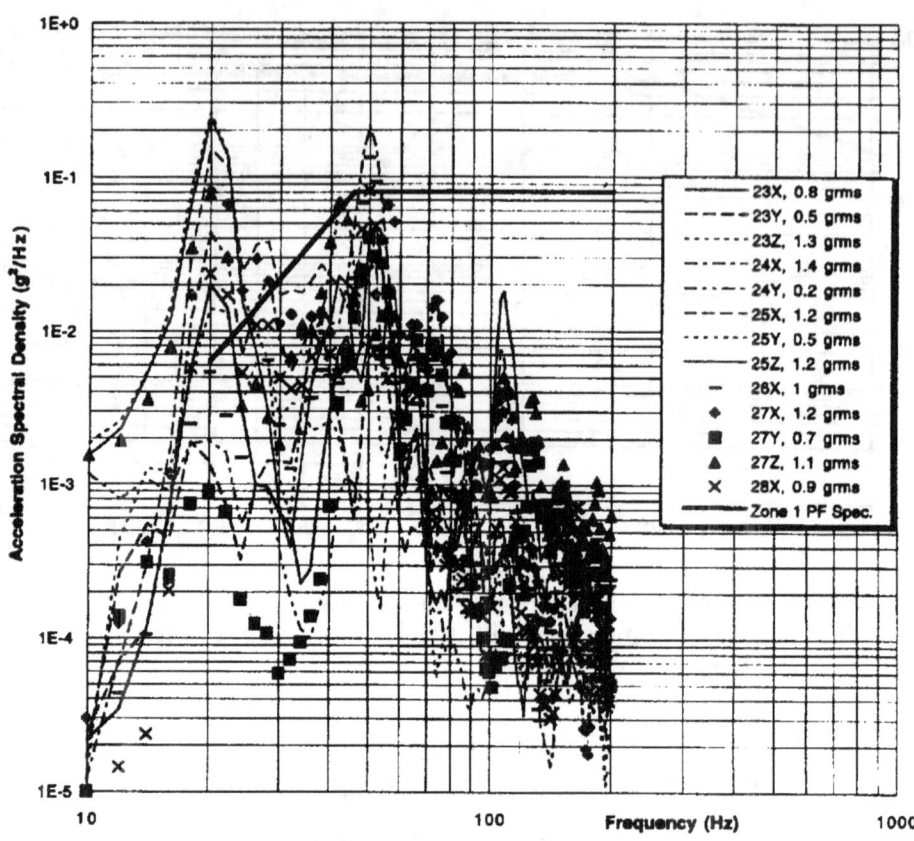

FIGURE 50. Acceleration Inputs to Remote Sensing Pallet Instruments in Cassini Full-Level Random Vibration Test (Comparison with Instrument Random Vibration Test Specification)

7.0 Conclusions

Here the key points discussed in this monograph are reiterated, and some suggestions for further development of the technology are offered.

1. Technical literature reviewed supports the conclusion that the pioneers of aerospace vibration testing recognized the dangers of vibration overtesting and understood that the very high shaker impedance was the culprit [1,2]. Many researchers during the subsequent thirty years studied the mounting structure impedance problem and developed conceptual solutions, many of which are a part of the force limiting approach described herein [4,10]. One can only conclude that either the instrumentation or the need was not sufficient to drive the implementation of this technology until recently. It is the author's belief that the recent advent of piezo-electric, tri-axial force transducers was the enabling factor.

2. In built-up aerospace configurations, the structural impedance of the mounting structure and test item are comparable and there is little amplification at the resonance frequencies of the test item. The interface acceleration has notches at the test item resonance frequencies, due to the vibration absorber effect.

3. The shaker, by contrast, has very high mechanical impedance and the test item can have very large amplification at its resonance frequencies. In addition, the test item resonance frequencies on the shaker occur at significantly higher frequencies than the coupled system resonances in the built-up configuration.

4. The object of force limited vibration testing is to make the input force at the test item resonances in the vibration test equal to the maximum interface force in the flight configuration. The goal is to replicate the internal forces and stresses in the flight environment, but the simulation is not exact because the resonance frequencies and mode shapes are different. The result of limiting the input force at resonances in the vibration test is that the input acceleration is notched in a manner similar to that due to the vibration absorber effect in flight. However, the notch frequencies will be slightly different on the shaker and in flight, due to the differences in off-axis boundary conditions.

5. The effective mass concept, developed in the early 70's [15], provides a theoretical basis for analyzing structural impedance data and FEM results. Until recently, the effective mass concept was know and used by only a select group of analysts, but the concept is gaining acceptance and it is hoped that an increased acceptance of the concept will be facilitated by the discussion in Section 3.3.4, provided by one of the concept's originators, as to the derivation of the effective mass from NASTRAN of other FEMs.

6. The philosophy adopted in the derivation of force specifications is to start with the traditional acceleration specification, which is the envelope of the flight interface acceleration, and add an analogous force specification, which is the envelope of the flight interface forces. Thus the force specifications derived analytically herein are proportional to the assumed acceleration specification, and any errors or undue conservatism in the acceleration specification carry over into the force specification. Thus force limiting should not be perceived as a method of compensating for errors in the acceleration specification. Rather, it is a method of automatically inserting notches in the acceleration spectrum at the proper frequencies and of the proper depth.

7. The general method of deriving force specifications is to develop a coupled system model of the source and load, with the modal parameters of each determined from FEM analysis or from impedance measurements. Then the ratio of the frequency <u>envelope</u> of the interface force to the <u>envelope</u> of the interface acceleration is determined. Finally, this ratio is multiplied by the acceleration specification, developed in a conventional manner (usually semi-empirically), to obtain the force specification. Two applications of this general method involving a simple and a complex TDFS are derived herein and the results are presented parametrically [27]. Other applications of the method are available in the literature [24] and it is envisioned that more sophisticated models will be developed in the future.

8. The literature [3,10,14] and spacecraft system acoustic tests [28] have provided data for the development of semi-empirical force limits which are much simpler to apply and appear to yield satisfactory results. Little flight data are currently available for vibratory force, but several flight measurement programs are in progress.

9. The advent of tri-axial, piezo-electric force transducers and of a new generation of digital controllers have facilitated the application of force limited vibration testing. The piezo-electric force transducers are easy-to-use, rugged, compact, have a wide dynamic range, and can readily be configured to measure all six components of force and moment [16]. The convenient measurement of total external force in a vibration test now makes it possible to measure the acceleration of the center-of-gravity, so that the design capability of aerospace structures can be conveniently verified in a vibration test.

10. There are many topics in the development of force limited vibration testing which require further investigation. Impedance methods are most convenient when two systems are connected at a single node, and this assumption is implicit herein. Many difficulties occur when one considers multi-point connections [8], which are almost always the case in the real world. The issue of overtesting due to uncorrelated inputs at multiple attachment points has not been addressed herein. Future vibration controllers will probably offer the capability to control phase, and then appropriate phase specifications between force and acceleration will have to be developed [4, 13]. New force transducers are being developed, including devices which can both measure and generate force, and new instrumentation developments will open the door to new testing techniques. Comparison of flight data with the prediction methods discussed herein will almost certainly give rise to some discrepancies, which will need to be resolved.

Appendix A Definition of Symbols

A	= interface acceleration
A_b	= base acceleration
A_o	= free acceleration of source
A_s	= acceleration specification
C	= dashpot constant
C	= constant
F	= interface force
F_o	= blocked force of source
F_s	= force specification or limit
F_e	= excitation force
k	= spring stiffness
k	= physical stiffness matrix
M_o	= total mass
M	= residual mass
m	= modal mass
$\underline{M}$	= apparent mass, F/A
m	= physical mass matrix
M	= modal mass matrix
Q	= dynamic amplification factor
S_{AA}	= acceleration spectral density
S_{FF}	= force spectral density
u	= absolute displacements
U	= generalized modal displacement
α	= ratio of modal to residual masses
β	= ratio of analysis frequency to resonance frequency
ϕ	= mode shape
μ	= ratio of load to source residual masses
ω	= radian frequency
ω_o	= natural frequency of uncoupled oscillator
Ω	= ratio of load to source uncoupled resonance frequencies

Subscripts

1	= source oscillator
2	= load oscillator
F	= unrestrained (free)
P	= prescribed
N	= modal set
R	= rigid body set
n	= single mode
p	= reaction force direction
q	= prescribed acceleration direction

Appendix B   Bibliography

1. Blake, R. E. , "The Need to Control the Output Impedance of Vibration and Shock Machines", Shock and Vibration Bulletin, No. 23, June 1956, pp. 59–64.

2. Morrow, C. T., "Application of Mechanical Impedance Concept to Shock and Vibration Testing", Los Angeles, CA, TRW Report AD 608030, 1960.

3. Salter, J. P. , "Taming the General-Purpose Vibration Test", Shock and Vibration Bulletin, No. 33, Pt. 3, March 1964, pp. 211–217.

4. Ratz, A. G., "An Impedance Compensated Random Equalizer", Proceedings of the Institute of Environmental Sciences 12th Annual Technical Meeting, San Diego, April 1966, pp.353–357.

5. Heinricks, J. A., "Feasibility of Force-Controlled Spacecraft Vibration Testing Using Notched Random Test Spectra", Shock and Vibration Bulletin, No. 36, Pt. 3, Jan. 1967.

6. McCaa, R. and Matrullo, M. "Flight Level Vibration Testing of a Lifting Body Re-entry Vehicle", Shock and Vibration Bulletin, No. 36, Pt. 3, Jan. 1967.

7. Painter, G. W., "Use of Force and Acceleration Measurements in Specifying and Monitoring Laboratory Vibration Tests", Shock and Vibration Bulletin, No. 36, Pt. 3, Jan. 1967.

8. O'Hara, G. J., "Mechanical Impedance And Mobility Concepts", J. of the Acoust. Soc. Am. 41(5), 1967.

9. Rubin, "Mechanical Immitance and Transmission-matrix Concepts", J. Acoust. Soc. Am. 41(5), 1967.

10. Murfin, W. B., "Dual Specification In Vibration Testing", Shock and Vibration Bulletin, No. 38, Pt. 1, 1968.

11. Scharton, T. D., "Development Of Impedance Simulation Fixtures For Spacecraft Vibration Tests", Shock and Vibration Bulletin, No. 40, Pt. 3, 1969, pp. 230–256.

12. Witte, A. F., "Specification Of Sine Vibration Test Levels Using A Force-Acceleration Product Technique", Shock and Vibration Bulletin, No. 41, Pt. 4, 1970, pp. 69–78.

13. Witte, A. F. and Rodeman, R., "Dual Specification In Random Vibration Testing, An Application Of Mechanical Impedance", Shock and Vibration Bulletin, No. 41, Pt. 44, 1970, pp. 109–118.

14. Hunter, N.F. and J. V. Otts, "The Measurement Of Mechanical Impedance And Its Use In Vibration Testing", Shock and Vibration Bulletin, No. 42, Pt. 1, 1972, pp. 79–88.

15. Wada, B. K., R. Bamford, and J. A. Garba, "Equivalent Spring-Mass System: A Physical Interpretation", Shock and Vibration Bulletin, No. 42, Pt. 5, 1972, pp. 215–225.

16. Martini, . K. H., "Multicomponent Dynamometers Using Quartz Crystals as Sensing Elements", ISA Transactions, Vol. 22, No. 1, 1983.

17. Judkins, N. J. and S. M. Ranaudo, "Component Internal Vibration Response Accelerations—System Level Versus Component Level", Proceedings of the 10th Aerospace Testing Seminar, 1987, pp. 97–104.

18. Sweitzer, K. A., "A Mechanical Impedance Correction Technique For Vibration Tests", Proceedings of the Institute of Environmental Sciences 33rd Annual Technical Meeting, 1987, pp. 73–76.

19. Piersol, A. G. , P. H. White, J. F. Wilby, P. J. Hipol, and E. G. Wilby, "Vibration Test Procedures For Orbiter Sidewall-Mounted Payloads: Phase I Report Appendices", Astron Research and Engineering (Santa Monica, CA), Rept. 7114-01, 1988.

20. Scharton, T. D. and D. L. Kern, "Using The VAPEPS Program To Support The TOPEX Spacecraft Design Effort", Proceedings of the 59th Shock and Vibration Symposium, 1988, pp. 21–36.

21. Scharton, T. D., Boatman, D. J. and Kern, D. L., "Dual Control Vibration Testing. Proceedings of the 60th Shock and Vibration Symposium",1989, pp.199–217.

22. Smallwood, D. O., "An Analytical Study Of A Vibration Test Method Using Extremal Control Of Acceleration And Force", Proceedings of the Institute of Environmental Sciences 35th Annual Technical Meeting, 1989, pp. 263–271.

23. Scharton, T. D., "Analysis Of Dual Control Vibration Testing", Proceedings of the Institute of Environmental Sciences 36th Annual Technical Meeting, 1990, pp. 140–146.

24. Smallwood, D. O., "Development Of The Force Envelope For An Acceleration/Force Controlled Vibration Test", Proceedings of the 61st Shock and Vibration Symposium, 1990, pp. 95–104.

25. Scharton, T. D., "Force Limited Vibration Testing at JPL", Proceedings of the Institute of Environmental Sciences 14th Aerospace Testing Seminar, 1993, pp. 241-251.

26. Scharton, T. D., "Force Limits for Vibration Tests", CNES Conference on Spacecraft Structures and Mechanical Testing, Paris, FR , June 1994, p. 1024.

27. Scharton, T. D., "Vibration-Test Force Limits Derived From Frequency-Shift Method", AIAA Journal of Spacecraft and Rockets 32(2), 1995, pp. 312–316.

28. Scharton, T.D., and Chang, K., "Force Limited Vibration Testing of the Cassini Spacecraft and Instruments", IES 17th Aerospace Testing Seminar, Los Angeles, Ca., Oct. 14, 1997.

29. Boatman, D., Scharton, T., Hershfeld, D., and Larkin, P., "Vibration and Acoustic Testing of the TOPEX/POSEIDON Satellite", Sound and Vibration Journal, Nov. 1992, pp. 14-16.

30. Den Hartog, J. P., *Mechanical Vibrations*, 3rd Ed., Mcgraw-Hill, New York, 1947, p. 115.

31. Neubert, V. H. , *Mechanical Impedance: Modeling/Analysis of Structures*, Josten Printing and Publishing, State College, Pa., 1987, pp.74-81.

32. Ewins D. J., *Modal Testing Theory and Practice*, John Wiley, New York, NY, 1984, p. 26.

33. Scharton, T.D., "Frequency-Averaged Power Flow into a One-Dimensional Acoustic System", J. Acoust. Soc. Am., 50 (1), 1971, p. 374.

34. Bamford, R. M., B. K. Wada, and W. H. Gayman, "Equivalent Spring-Mass System For Normal Modes", Jet Propulsion Laboratory (Pasadena, CA), Technical Memorandum 33-3803, 1971.

35. Skudrzyk, E., *Simple and Complex Vibratory Systems*, Pennsylvania State University Press, University Park, PA, 1968.

36. Skudrzyk, E., "The Mean-value Method of Predicting the Dynamic Response of Complex Systems", J. Acoust. Soc. Am., 67 (4), April 1980, p.1105.

37. Scharton, T. D., "Vibration Test Force Limits Derived from Frequency Shift Method", Jet Propulsion Laboratory, Pasadena, CA, JPL D-11455, Jan. 1994.

38. Crandall, S. H. and W. D. Mark, *Random Vibration in Mechanical Systems*, New York: Academic Press, 1973.

39. Chang, K and Scharton, T.D., "Verification of Force and Acceleration Specifications for Random Vibration Tests of Cassini Spacecraft Equipment", Proc. of ESA Conference on Spacecraft Structures, Materials & Mechanical Testing, Netherlands, 27-29 March, 1996.

40. Trubert, M., "Mass Acceleration Curve for Spacecraft Structural Design", NASA Jet Propulsion Laboratory, JPL D-5882, Nov. 1, 1989.

41. Smith, K., "Equations for JPL Loads Analysis", NASA Jet Propulsion Laboratory, JPL D-6018, Oct. 1, 1990.

42. Scharton, T. D., "Force-Limited Vibration Tests at JPL--A Perfect Ten", ITEA Jour. of Test and Evaluation, Vol. 14, No. 3, September 1993.

43. Rentz, P.R., "Cassini Flight Spacecraft Protoflight Random Vibration Test Report," JPL-D-14198, Jet Propulsion Laboratory, Pasadena, CA, 24 March 1997.

Appendix C   Force Limited Vibration Testing Handbook (Draft)

This draft, dated 04/09/97 prepared by T. Scharton / JPL, has not been approved and is subject to modification. DO NOT USE PRIOR TO APPROVAL.

**National Aeronautics and
Space Administration**

MEASUREMENT
SYSTEM ID

April 1997

FORCE LIMITED
VIBRATION TESTING

NASA TECHNICAL HANDBOOK

FORWARD

This handbook is approved for use by NASA Headquarters and all field centers and is intended to provide a common framework for consistent practices across NASA programs.

The primary goal of vibration tests of aerospace hardware is to identify problems which, if not remedied, would result in flight failures. This goal can only be met by implementing a realistic (flight-like) test with a specified positive margin. In most cases, the goal is not well served by traditional acceleration controlled vibration tests which indeed screen out flight failures but in addition, cause failures, which would not occur in flight. This overtest or "screening" test approach, which may have served its purpose in the past, is too expensive and inefficient for today's environment of low-cost missions. The penalty of overtesting is manifested in design and performance compromises, as well as in the high costs and schedule overruns associated with recovering from artificial test failures.

It has been known for thirty years that the major cause of overtesting in aerospace vibration tests is associated with the infinite mechanical impedance of the shaker and the standard practice of controlling the input acceleration to the frequency envelope of the flight data. This approach results in artificially high shaker forces and responses at the resonance frequencies of the test item. To alleviate this problem it has become common practice to notch the input acceleration to limit the responses in the test to those predicted for flight, but this approach is very dependent on the analysis, which the test is supposed to validate. Another difficulty with response limiting is that it requires placing accelerometers on the test item at all the critical locations, many of which are often inaccessible, and which in the case of large test items, involves extensive instrumentation.

The advent of new instrumentation has made possible an alternative, improved vibration testing approach based on measuring and limiting the reaction force between the shaker and test item. The major break through is the availability of piezo-electric triaxial force gages developed for other commercial markets. Piezo-electric force gages are robust, relatively easy to install between the test item and shaker, and require the same signal conditioning as piezo-electric accelerometers commonly used in vibration testing. Also, a new generation of vibration test controllers now provide the capability to limit the measured forces and thereby notch the input acceleration, in real time. To take advantage of this new capability to measure and control shaker forces, a rationale for predicting the flight limit forces has been developed and applied to many flight projects during the past five years. Force limited vibration tests are now conducted routinely at the Jet Propulsion Laboratory (JPL) and also at several other NASA centers, government laboratories, and many aerospace contractors.

This handbook describes an approach which may be used to facilitate and maximize the benefits of applying this relatively new technology throughout NASA in a consistent manner. A monograph, which provides more detailed information on the same subject, is also available for reference.

Requests for information, corrections, or additions to this handbook should be directed to the Mechanical Systems Engineering and Research Division, Section 352, Jet Propulsion Laboratory, 4800 Oak Grove Dr., Pasadena, CA 91109. Requests for additional copies of this handbook should be sent to NASA Engineering Standards, EL02, MSFC, AL, 35812 (telephone 205-544-2448).

Daniel R. Mulville
Chief Engineer

CONTENTS

iii

1.1   Purpose

This handbook establishes a methodology for conducting force limited vibration tests for all NASA flight projects. The purpose is to provide an approach which may be consistently followed by those desiring to use force limiting, without having to conduct an extensive literature search or research and development effort before conducting the test. The decision to use, or not use, force limiting on a specific project and in a specific vibration test, and the responsibility for applying the method correctly, are left to the project or the cognizant engineer. A monograph on Force Limited Vibration Testing is available for reference and it is recommended for those needing more detailed technical information.

1.2   Applicability

This handbook recommends engineering practices for NASA programs and projects. It may be cited in contracts and program documents as a technical requirement or as a reference for guidance. Determining the suitability of this handbook and its provisions is the responsibility of program/project management and the performing organization. Individual provisions of this handbook may be tailored (i.e., modified or deleted) by contract or program specifications to meet specific program/project needs and constraints.

For the purpose of this handbook, a force limited vibration test is any vibration test in which the force between the test item and shaker is measured and controlled. (The recommended means of measuring the force is with piezo-electric force gages, but other means, e.g. shaker armature current or strain gages, may be useful in special situations. Similarly, the control of the force is preferably accomplished in real time, but iterative, off-line control may be employed as a stepping stone.) If the force is not measured and controlled, the test is not considered a force limited vibration test, and this handbook does not apply. This distinction is important because in the past some have found it convenient to simulate a force limited test and then to use the analytical results to notch the acceleration input in the test. The simulation approach is not recommended because measurement of the force is considered to be the essential element of the force limiting approach.

The handbook is applicable to all force limited vibration tests of NASA flight hardware including launch vehicle and aircraft equipment, spacecraft, instruments, and components. However since the purpose of force limiting is to mitigate the effect of test item resonances in the vibration test, the technique is most useful for structure-like equipment and for fragile equipment such as optics and complex instruments.

2.0 APPLICABLE DOCUMENTS

2.1 General. The applicable documents cited in this handbook are listed in this section only for reference. The specified technical requirements listed in the body of this document must be met whether or not the source document is listed in this section.

2.2 Government documents. The following Government documents form a part of this document to the extent specified herein. Unless otherwise specified, the issuances in effect on date of invitation for bids or request for proposals shall apply.

NASA - RP-1403   -   Force Limited Vibration Test
                                                                Monograph, May 1997

NASA - GUIDELINES - XXXX   -   Dynamic Environment
                                                                Guidelines, September 1997

1

NASA - STD - 7001   -   Payload Vibroacoustic
Test Criteria, June 21, 1996.

NASA - STD - 7002   -   Payload Test Requirements,
July 10, 1996.

2.3 <u>Non-government publications</u>. The following documents a part of this document to the extent specified herein. Unless otherwise specified, the issuances in effect on the date of invitation for bids or request for proposals shall apply.

Blake R. E., "The Need to Control the Output Impedance of Vibration and Shock Machines", Shock and Vibration and Associated Environments, Bulletin No. 23, 1954.

Salter, J. P., "Taming the General-Purpose Vibration Test", Shock and Vibration and Associated Environments, Bulletin No. 33, Part III, March 1964, pp. 211-217.

Murfin, W. B., "Dual Specifications in Vibration Testing", Shock and Vibration Bulletin, No. 38., Part 1, 1968, pp. 109-113.

Wada, B. K., Bamford, R., and Garba, J. A., "Equivalent Spring-Mass System: A Physical Interpretation", Shock and Vibration Bulletin, No. 42, 1972, pp. 215-225.

Scharton, T. D., Boatman, D. J., and Kern, D. L., "Dual Control Vibration Testing", Proceedings of 60th Shock and Vibration Symposium, Vol. IV, 1989, pp. 199-217.

Smallwood, D. O., "An Analytical Study of a Vibration Test Method Using Extremal Control of Acceleration and Force", Proceedings of Institute of Environmental Sciences 35th Annual Technical Meeting, 1989, pp. 263-271.

Scharton, T. D., "Analysis of Dual Control Vibration Testing", Proceedings of Institute of Environmental Sciences 36th Annual Technical Meeting, 1990, pp. 140-146.

Smallwood, D. O., "Development of the Force Envelope for an Acceleration/Force Extremal Controlled Vibration Test", Proceedings of 61st Shock and Vibration Symposium, Vol. I, 1990, pp. 95-104.

Scharton, T. D., "Force Limited Vibration Testing at JPL", Proceedings of the Institute of Environmental Sciences 14th Aerospace Testing Seminar, 1993, pp. 241-251.

Scharton, T. D., "Vibration-Test Force Limits Derived from Frequency-Shift Method", AIAA Journal of Spacecraft and Rockets, Vol. 2, No. 2, March 1995, pp. 312-316.

Scharton, T., Bamford, R., and Hendrickson, J., "Force Limiting Research and Development at JPL", Spacecraft and Launch Vehicle Technical Information Meeting, Aerospace Corp., Los Angeles, CA, June 7, 1995.

Chang, K. Y. and Scharton, T. D., "Verification of Force and Acceleration Specifications for Random Vibration Tests of Cassini Spacecraft Equipment", ESA/CNES Conference on Spacecraft Structures, Materials, and Mechanical Testing, Noordwijk, NL, March 27-29, 1996.

2.4 <u>Order of precedence.</u> Where this document is adopted or imposed by contract on a program or project, the technical guidelines of this document take precedence, in the case of

2

conflict, over the technical guidelines cited in other referenced documents. This handbook does not apply to payload programs approved prior to the date of this document. Also, this handbook does not address safety considerations that are covered thoroughly in other documents; but if a conflict arises, safety shall always take precedence. Nothing in this document, however, supersedes applicable laws and regulations unless a specific exemption has been obtained.

3.0 DEFINITIONS

ACCELERANCE--Complex frequency response function which is ratio of acceleration to force

ACCELERATION OF C.G.--Acceleration of instantaneous centroid of distributed masses (equal to external force divided by total mass, according to Newton's 2nd Law)

APPARENT MASS--Complex frequency response function which is ratio of force to acceleration

CONTROL SYSTEM--The hardware and software which provides means for the test operator to translate vibration specifications into the drive signal for the shaker

DESIGN VERIFICATION TEST--Test to see if as-built test-item can survive design loads

DUAL CONTROL--Control of both force and vibration

DYNAMIC ABSORBER--SDFS tuned to excitation frequency to provide reaction force which reduces motion at attachment point

EFFECTIVE MASS-- Masses in model consisting of SDFS's connected in parallel to a common base, so as to represent the apparent mass of a base driven continuous system. The sum of the effective modal masses equals the total mass.

EXTREMAL CONTROL--A shaker controller algorithm based on control of the maximum (extreme) of a number of inputs in each frequency control band

FLIGHT LIMITS--Definition of accelerations or forces which are believed to be equal to the maximum flight environment, often P(95/50)

FORCE LIMITING--Reduction of the reaction forces in a vibration test to specified values, usually to the interface forces predicted for flight, plus a desired margin

IMPEDANCE--Complex frequency response function which is ratio of force to velocity quantities (Sometimes used to refer to ratio of force to any motion quantity.)

LEVEL--Test input or response in decibels (dB), dB = 20 log amplitude = 10 log power

LOAD--Vibration test item

MARGIN--Factor to be multiplied times, or decibels to be added to, the flight limits to obtain the test specification

NOTCHING--Reduction of acceleration input spectrum in narrow frequency bands, usually where test item has resonances

QUALITY FACTOR--Measure of the amplification of the response at resonance, equal to the reciprocal of twice the critical damping ratio.

QUASI-STATIC ACCELERATION--Combination of static and low frequency loads into an equivalent static load specified for design purposes as C.G. acceleration

RESIDUAL MASS-- Sum of the effective masses of all the vibration modes with resonance frequencies greater than the excitation frequency.

RESPONSE LIMITING-- Reduction of input acceleration to maintain measured response at or below specified value, usually as predicted for flight plus desired margin

SHAKER--The machine which provides vibratory motion to the test item, usually electrodynamic, in aerospace testing (can also be hydraulic or rotary)

THREE AXIS LOAD CELL--Force gage which measures the three perpendicular components of force simultaneously

SINGLE DEGREE-OF-FREEDOM SYSTEM (SDFS) --Vibration model with one mass

SOURCE--Test item support structure which provides vibration excitation inflight

SPECIFICATIONS--Definition of vibration quantity versus frequency, usually associated with programmatic requirements

TAP TEST--Measurement of apparent mass or accelerance by tapping on structure with small rubber or plastic tipped hammer which incorporates force transducer

TEST FIXTURE--Adapter hardware which allows test item to be mounted to shaker

TWO DEGREE-OF-FREEDOM SYSTEM (TDFS) --Vibration model with two masses

4.0 GENERAL REQUIREMENTS

4.1 Criteria for Force Limiting. The purpose of force limiting is to reduce the response of the test item at its resonances on the shaker in order to replicate the response at the combined system resonances in the flight mounting configuration. Force limiting is most useful for structure-like test items which exhibit distinct, lightly damped resonances on the shaker. Examples are complete spacecraft, cantilevered structures like telescopes and antennas, lightly damped assemblies such as cold stages, fragile optical components, and equipment with pronounced fundamental modes such as a rigid structure with flexible feet. The amount of relief available from force limiting is greatest when the structural impedance (effective mass) of the test item is equal to, or greater than, that of the mounting structure. However, it is recommended that notches deeper than 14 dB not be implemented without appropriate peer review. Force limiting is most beneficial when the penalties of an artificial test failure are high. Sometimes this is after an initial test failure in a screening type of test.

4.2 Instrumentation.

4.2.1 Piezo-electric Force Gages. The use of piezo-electric force gages for force limiting is highly recommended over other types of force measurement means such as strain gages, armature current., etc. The advent of piezo-electric, quartz force gages has made the measurement of force in vibration tests almost as convenient and accurate as the

4

measurement of acceleration. The high degree of linearity, dynamic range, rigidity, and stability of quartz make it an excellent transducer material for both accelerometers and force gages. Similar signal processing, charge amplifiers and voltage amplifiers, may be used for piezo-electric force gages and accelerometers. However, there are several important differences between these two types of measurement. Force gages must be inserted between (in series with) the test item and shaker and therefore require special fixturing, whereas accelerometers are placed upon (in parallel with) the test item or shaker. The total force into the test item from several gages placed at each shaker attachment may be obtained by simply using a junction to add the charges before they are converted to voltage, whereas the output of several accelerometers is typically averaged rather than summed. Finally, piezo-electric force gages tend to put out more charge than piezo-electric accelerometers because the force gage crystals experience higher loading forces, so sometimes it is necessary to use a charge attenuator before the charge amplifier.

4.2.2 <u>Force Gage Preload</u>. Piezo-electric force gages must be preloaded so that the transducer always operates in compression. Having a high preload and smooth transducer and mating surfaces minimizes several common types of gage measurement errors, e.g. bending moments being falsely sensed as tension/compression. However, using flight hardware and fasteners, it is usually impossible to achieve the manufacturers recommended preload. In addition, sometimes it is necessary to trade-off transducer preload and dynamic load, particularly moment, carrying capability. The two requirements for selecting the preload are that it is sufficient to prevent unloading due to the dynamic forces and moments and that the maximum stress on the transducers does not exceed that associated with the manufacturer's recommended maximum load configuration.

Transducer preloading is applied using a threaded bolt or stud which passes through the inside diameter of the transducer. With this installation, the bolt or stud acts to shunt past the transducer a small portion of any subsequently applied load, thereby effectively reducing the transducer's sensitivity. Calibration data for the installed transducers is available from the manufacturer it they are installed with the manufacturer's standard mounting hardware. Otherwise, the transducers must be calibrated in situ.

4.2.3 <u>Force Gage Calibration.</u> The force gage manufacturer provides a nominal calibration for each transducer, but the sensitivity of installed units must be determined in situ, as discussed in the previous paragraph. This may be accomplished either quasi-statically or dynamically. Using the transducer manufacturer's charge amplifiers and a low noise cable, the transducers will hold their charge for several hours, so it is possible to calibrate them statically with weights or with a hydraulic loading machine. It is recommended that the calibration be performed by loading the transducers, zeroing out the charge, and then removing the load, in order to minimize the transient overshoot.

The simplest method of calibrating the transducers for a force limited vibration test is to conduct a preliminary low level sine sweep or random run and to compare the apparent mass (ratio of total force in the shaker direction to the input acceleration) measured at frequencies much lower than the first resonance frequency with the total mass of the test item. Typically the measured force will be approximately 80 to 90 % of the weight in the axial direction and 90 to 95% of the weight in the lateral directions, where the preloading bolts are in bending rather than in tension or compression. Alternately, the calibration correction factor due to the transducer preloading bolt load path may be calculated by partitioning the load through the two parallel load paths according to their stiffness; the transducer stiffness is provided by the manufacturer, and the preload bolt stiffness in tension and compression or bending must be calculated.

4.2.4   Force Gage Combinations   It is recommended that the total force in the shaker excitation direction be measured in a force limited vibration test. The total force from a number of gages in parallel is readily obtained using a junction box which sums the charges, and therefore the forces, before conditioning the signal with a charge amplifier. An alternative is to specify limits for the force at individual attachment positions, but this is not recommended. Since vibration tests are normally conducted sequentially in three perpendicular axes, it is convenient to employ triaxial force transducers. Sometimes it is necessary to limit the cross-axis force and the moments in addition to the in-axis force; this is particularly the case in tests of large eccentric test items such as spacecraft. For these applications, the six force resultant forces and moments for a single node may be measured with a combination, commonly four, of triaxial force transducers and a voltage summer.

4.2.5   Accelerometers. Accelerometers on the fixture are also required in force limited vibration tests in order to control the acceleration input to the acceleration specification at frequencies other than at the test item resonances. In addition, it is often convenient to use a limited number of accelerometers to measure the response at critical positions on the test item. These response accelerometers may be used only for monitoring or, if justified by appropriate rationale, for response limiting in addition to the force limiting.

4.3 Fixturing. The preferred method of configuring the force gages is to sandwich one gage between the test item and conventional test fixture at each attachment position and use fasteners which are longer than the conventional ones to accommodate the height of the gages. In this configuration, there is no fixture weight above the transducers and the gage force is identical to the force into the test item. Sometimes the preferred approach is impractical, e. g. if there are too many attachment points or the attachments involve shear pins in addition to bolts. In these cases it may be necessary to use one or more adapter plates to interface the transducers to the test item. The requirement is that the total weight of the adapter plates above the force gages does not exceed ten percent of the weight of the test item. This limitation is necessary because the force gages read the sum of the force required to accelerate the interface plate and that delivered to the test item. If the fixture weight exceeds the 10% criterion, force limiting will only be useful for the first one or two modes in each axis. Use of a circuit to subtract the interface plate force in real time, is not recommended because of the errors that result when the interface plate is not rigid. The use of armature current to measure shaker force is also not generally useful, because the weight of the armature and fixturing typically are much greater than 10 % of that of the test item.

4.4 Force Specifications. Force limits are analogous and complementary to the acceleration specifications used in conventional vibration testing. Just as the acceleration specification is the frequency spectrum envelope of the inflight acceleration at the interface between the test item and flight mounting structure, the force limit is the envelope of the inflight force at the interface. In force limited vibration tests, both the acceleration and force specifications are needed, and the force specification is proportional to the acceleration specification. Therefore force limiting does not compensate for errors in the development of the acceleration specification, e.g. undue conservatism or lack thereof. These errors will carry over into the force specification. Since inflight vibratory force data are lacking, force limits are usually derived from coupled system analyses and impedance information obtained from measurements or finite element models (FEM). Fortunately, considerable data on the interface force between spacecraft and components are becoming available from spacecraft acoustic tests, and semi-empirical methods of predicting force limits are being developed.

4.4.1 Analytical Force Limits. Analytical models and methods of obtaining impedance information to use in these models are discussed in Section 5.0 Detailed Requirements. Here, the general requirements for analytical force limits are discussed. It is required that analytical models used to predict force limits take into account the resonant behavior of both

6

the source (mounting structure) and the load (test item) and that the models incorporate impedance information, data or FEM, on both the source and the load. The models discussed in the Detailed Requirements section are two-degree-of-freedom system (TDFS) models, in which the coupled source and load are each described by a single resonant mode. In more complex models, the source and load may have many modes. In the early stages of a program, before hardware exists, strength of materials or FEM models are often used to determine the modal parameters of the source and load. Later in the program, before the vibration tests of flight hardware, it is recommended that the modal parameters be updated with impedance data measured in tap tests on the mounting structure and in the shaker tests of the test item. The coupled source and load models are exercised with some representative excitation of the source, and the envelope (or peak values) of the interface acceleration and interface force frequency response functions (FRF) are calculated, preferably in one-third octave bands. Finally, the ratio of the interface force envelope to the acceleration envelope is calculated from the model and the force limit specification is calculated by multiplying the conventional acceleration specification by this ratio. (It is essential that the ratio of envelopes or peaks, not of the FRF's, be calculated.)

4.4.2 <u>Semi-empirical Force Limits</u>. The alternative semi-empirical approach to deriving force limits is based on the extrapolation of interface force data for similar mounting structure and test items. A general form for a semi-empirical force limit for sine or transient tests is from "Taming the General-Purpose Vibration Test":

$$F_s = C \ M_o \ A_s \tag{1a}$$

where F_1 is the amplitude of the force limit, C is a frequency dependent constant which depends on the configuration, M_o is the total mass of the load (test item), and A_s is the amplitude of the acceleration specification. The form of Eq. 1a appropriate for random vibration tests is:

$$S_{FF} = C^2 \ M_o^2 \ S_{AA} \tag{1b}$$

where S_{FF} is the force spectral density and S_{AA} the acceleration spectral density.

As shown in "Verification of Force and Acceleration Specifications for Random Vibration Tests of Cassini Spacecraft Equipment", interface force data measured between three instruments, each weighing approximately 60 lb., and JPL's Cassini spacecraft in acoustic tests of the development test model spacecraft fit Eq. 1b with C equal to unity at frequencies up to and including the fundamental resonance of the test item and then with C rolling off as one over frequency at higher frequencies. It is required to show similarity between the subject hardware configuration and the reference data case or to justify the scaling used for any extrapolation, in order to use semi-empirical force limits.

4.4.3 <u>Quasi-static Design Verification</u>. The quasi-static design of aerospace components is often based on a specified acceleration of the center-of-gravity (C.G.) of the component. However, the C.G. of a flexible body is a virtual (not a real) point and its acceleration cannot be accurately measured with an accelerometer in a vibration test, particularly at frequencies above the fundamental resonance. However, Eq. 1a with C equal to unity and A_s equal to the C.G. acceleration is Newton's 2nd law. Thus limiting the external force to the product of total mass times the quasi-static design limit, or some fraction thereof, is the recommended method of validating quasi-static designs in vibration tests.

4.5 <u>Control System</u>. Most of the current generation of vibration test controllers have the two capabilities needed to implement force limiting. First, the controller must be capable of extremal control, sometimes called maximum or peak control by different vendors. In extremal control, the largest of a set of signals is limited to the reference spectrum. (This is in contrast to the average control mode in which the average of a set of signals is compared to the reference signal.) Most controllers used in aerospace testing laboratories support the extremal control mode. The second capability required is that the controller must support different reference spectra for the response limiting channels, so that the force signals may have limit criteria specified as a function of frequency. Controllers which support different reference spectra for limit channels are now available from most venders and in addition upgrade packages are available to retrofit some of the older controllers for this capability. If the controller does not have these capabilities, notching of the acceleration specification to limit the measured force to the force specification must be done manually in low level runs.

4.6 <u>Test Planning Considerations</u>. Several considerations need to be addressed in the test planning if force limiting is to be employed. First the size, number, and availability of the force transducers need to be identified as well as any special fixturing requirements to accommodate the transducers. Next, the approach for deriving and updating the force specification needs to be decided. Finally the control strategy, which in special cases may include cross-axis force, moment, individual force, and response limiting in addition to or in lieu of the in-axis force, must be decided and written into the test plan. In some instances, the control strategy will be limited by the control system capabilities. In all cases, it is recommended that the control strategy be kept as simple as possible, in order to expedite the test and to minimize the possibility of mistakes.

5.0 DETAILED REQUIREMENTS

5.1 <u>Derivation of Force Limits</u>. As the force limiting technology matures, there will eventually be as many methods of deriving force limits as there are of deriving acceleration specifications. Herein several acceptable methods are described.

Force spectra have typically been developed in one-third octave bands (see example in Section 6.2), but other bandwidths, e.g. octave or one-tenth octave bands, may also be used. Force limiting may usually be restricted to an upper frequency encompassing approximately the first three modes in each axis; which might be approximately 100 Hz for a large spacecraft, 500 Hz for an instrument, or 2000 Hz for a small component. It is important to take into account that the test item resonances on the shaker occur at considerably higher frequencies than in flight. Therefore care must be taken not to roll off the force specification at a frequency lower than the fundamental resonance on the shaker and not to roll off the specification too steeply, i.e. it is recommended that the roll-offs of the force spectrum be limited to approximately 9 dB/octave.

5.1.1 <u>Simple TDFS</u>. The simple Two-Degree-of-Freedom System (TDFS) method of deriving force limits is described in "Vibration-Test Force Limits Derived from Frequency-Shift Method". The basic model is shown in Figure 1. The model represents one vibration mode of the source (system 1) coupled with one vibration mode of the load (system 2). Figure 2 shows the ratio of the interface force spectral density S_{FF} to the input acceleration spectral density S_{AA}, normalized by the load mass M_2 squared, as a function of the mass ratio M_2/M_1, calculated from the simple TDFS. When this mass ratio is very small, there is no force limiting effect; the force spectral density asymptote is the load mass M_2 squared times the input acceleration spectral density times the quality factor Q_2 squared. The ratio of this asymptotic value of the force to the force limit at larger values of M_2/M_1, is the expected amount of notching, sometimes called the knock-down factor, when the force is limited to the force limit. The force limit is very insensitive to damping at values of M_2/M_1 greater than 0.4, but the unnotched force spectrum and therefore the notch depth resulting from force limiting will be proportional to the actual quality factor Q_2 squared. To use Figure 2, the source and load masses must be determined from FEM analyses or measurements as a function of frequency. It is recommended that one-third octave frequency bands be utilized. In the simple TDFS method, it is recommended for conservatism that these masses be taken as the residual masses rather than the modal masses. Appendix A gives the equations for replicating the curves in Figure 2.

5.1.2 <u>Complex TDFS</u>. The complex Two-Degree-of-Freedom System (TDFS) method of deriving force limits is also described in "Vibration-Test Force Limits Derived from Frequency-Shift Method". The complex TDFS model is shown in Figure 3; it requires both the modal (m) and the residual (M) masses of the source and load. Table 1 tabulates the normalized ratio of interface force spectral density to input acceleration spectral density for a complex TDFS with Q=20, which is a good nominal value for most practical applications. It is recommended that both the simple and complex TDFS models be used and that the larger of the two calculations be used in each one-third octave frequency band. It will generally be found that the simple TDFS gives the larger result off the load resonances and the complex TDFS the larger result at the load resonances.

5.1.3 <u>Multiple Degree-of-Freedom Systems</u>. In general, a multiple degree-of-freedom model of the source and load may be utilized as in "An Analytical Study of a Vibration Test Method Using Extremal Control of Acceleration and Force". The model parameters are determined from modal mass and resonance frequency information for the source and load. The ratio of the interface force envelope to the interface acceleration envelope should be evaluated as with simpler models, and the force limit determined by multiplying this ratio by the acceleration specification obtained as in conventional vibration tests.

5.1.4 <u>Alternative Methods</u>. Just as there are many ways of developing acceleration specifications, there will be many ways of deriving force limits. In time a data base of flight and system test force data and validated semi-empirical methods will be available, but for the present most force limits must be derived from analytical models with structural impedance data. Although the methods recommended in this handbook are preferred, other methods may be acceptable if they are rational and result in a desired margin over flight. One alternative method is to use the blocked force, which is the force that the source will deliver to an infinite impedance (zero motion) load. Unfortunately for most systems, the blocked force is too large to result in much limiting as shown in "Force Limiting Research and Development at JPL". Another method suitable for low frequency testing is to base the force limit on the C.G. acceleration from a mass-acceleration type curve such as is sometimes used for quasi-static design. See Section 4.4.3.

5.2. <u>Apparent and Effective Mass</u>. The frequency response function (FRF) which is the ratio of the reaction force to applied acceleration at the base of a structure is called "apparent mass". The apparent mass is a complex impedance-like quantity which reflects the mass, stiffness, and damping characteristics of the structure. The modal models recommended herein require only the "effective" masses, which are real quantities and therefore much simpler.

5.2.1. <u>Effective Mass Concept</u>. The concept of effective mass was introduced in "Equivalent Spring-Mass System: A Physical Interpretation". Consider the drive point apparent mass of the model consisting of the set of single-degree-of-freedom systems (SDFS) connected in parallel to a rigid, massless base as shown in Fig. 3, from "Vibration-Test Force Limits Derived from Frequency-Shift Method". The modal contribution to this

9

drive point apparent mass, divided by the SDFS frequency response factor, is called the effective mass of that mode. The sum of the effective modal masses is the total mass of the distributed system. The sum of the effective masses of the modes with resonance frequencies above the excitation frequency is called the effective residual mass. Appendix B provides a more general definition of effective mass and a procedure for using NASTRAN to calculate the effective masses.

5.2.2 <u>Shaker Measurement of Load Effective Mass</u>. The load effective residual mass should be measured and used to update the calculated force limits before conducting a force limited vibration test of flight hardware, because the force limits in both the simple and complex TDFS models are proportional to the load effective residual mass. Fortunately, the load effective residual mass can be readily measured with a low level sine sweep, or random, test run when the load is mounted with force gages on the shaker. First the magnitude of the drive point apparent mass, the ratio of total reaction force in the excitation direction to the input acceleration, is measured. Then this apparent mass function is smoothed (a moving average in frequency) to eliminate the resonance peaks. The resulting smooth curve, which must be a decreasing function of frequency by Foster's theorem, is taken as the effective residual mass. The effective mass for each distinguishable mode may be evaluated by equating the corresponding peak in the apparent mass curve to the sum of the residual mass and the product of the effective mass times the quality factor Q, determined from half-power bandwidth of the peak.

5.2.3 <u>Tap Test Measurement of Source Effective Mass</u>. The source effective residual mass is determined in a similar manner by smoothing the FRF's of the magnitude of the drive point apparent mass of the source, which are measured with a modal hammer incorporating a force gage. The measurements involve tapping at representative positions where the load attaches and computing the FRF of the ratio of the force to the acceleration, which is measured with an accelerometer mounted temporarily on the source structure near the hammer impact point. (The load must not however be attached to the source structure during these measurements.) Some judgment is involved in combining the apparent masses measured at multiple attachment points to obtain a single-node model of the effective mass. At low frequencies, each point yields the total mass, unless rotations are introduced. At high frequencies, the apparent masses from multiple points should be added, usually by adding the sum of the squares. Also when calculating or measuring the apparent mass of a mounting structure, it is important to decide how much of the adjacent structure it is necessary to consider. It is necessary to include only enough of the mounting structure so that the source effective modal and residual masses are accurately represented in the frequency range of the load resonances.

6.0 NOTES (This section is for information only and is not mandatory.)

6.1 <u>Reduction of Mean-Square Response Due to Notching</u>. It is often important to know how much the mean-square response, or force, will be reduced when a resonance is limited to some value. Limiting a response resonance to the peak spectral density divided by the factor A squared, results in a notch of depth A squared in the input spectral density at the response resonance frequency. The reduction in response resulting from notching is considerably less than that associated with reducing the input spectral density at all frequencies, in which case the response is reduced proportionally. The reduction in mean-square response of a SDFS resulting from notching the input dB = 20 log A at the response resonance frequency is shown in Fig. 4, from "Force Limiting Research and Development at JPL".

6.2 <u>Force Specification Example</u>. Appendix C is a spread sheet calculation of the force specification for an instrument (CRIS) mounted on a honeycomb panel of a spacecraft (ACE) using three methods: the simple TDFS, the complex TDFS, and the semi-empirical.

6.3 <u>Definition of Symbols</u>

A	= interface acceleration
Ab	= base acceleration
Ao	= free acceleration of source
As	= acceleration specification
C	= dashpot constant
C	= constant
F	= interface force
Fs	= force specification or limit
k	= spring stiffness
k	= physical stiffness matrix
M_o	= total mass
M	= residual mass
m	= modal mass
<u>M</u>	= apparent mass, F/A
m	= physical mass matrix
M	= modal mass matrix
Q	= dynamic amplification factor
S_{AA}	= acceleration spectral density
S_{FF}	= force spectral density
u	= absolute displacements
U	= generalized modal displacement
ϕ	= mode shape
ω	= radian frequency
ω_o	= natural frequency of uncoupled oscillator

Subscripts

1	= source oscillator
2	= load oscillator
F	= unrestrained (free)
P	= prescribed
N	= modal set
R	= rigid body set
n	= single mode
p	= reaction force direction
q	= prescribed acceleration direction

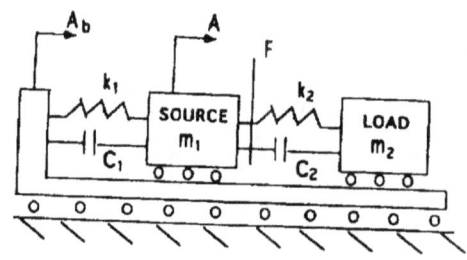

FIGURE 1. Simple TDFS of Coupled Oscillators

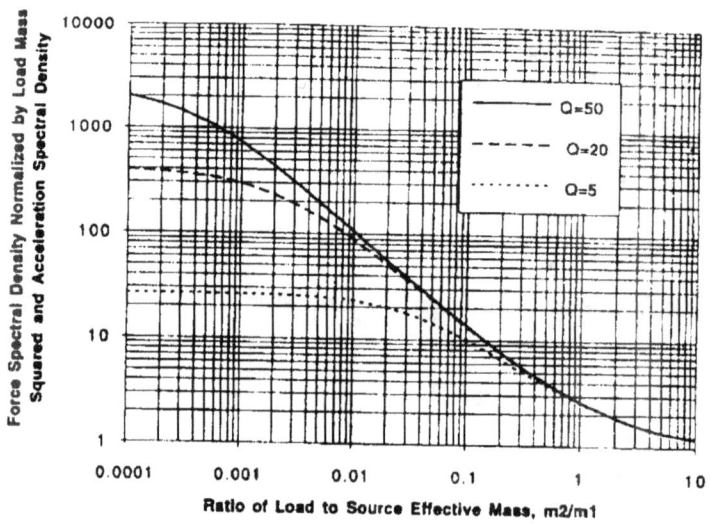

FIGURE 2. Normalized Force Spectrum for Simple TDFS

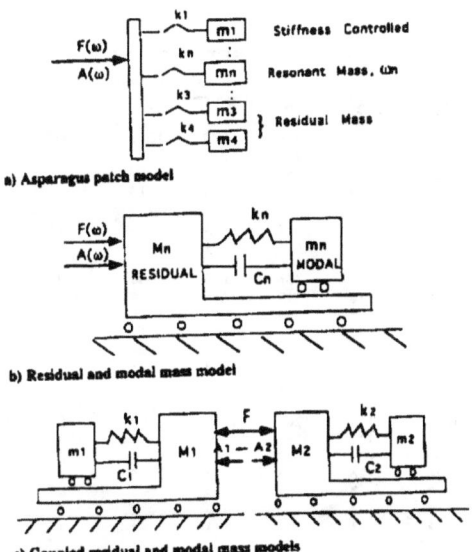

a) Asparagus patch model

b) Residual and modal mass model

c) Coupled residual and modal mass models

FIGURE 3. Complex TDFS with Residual and Modal Masses

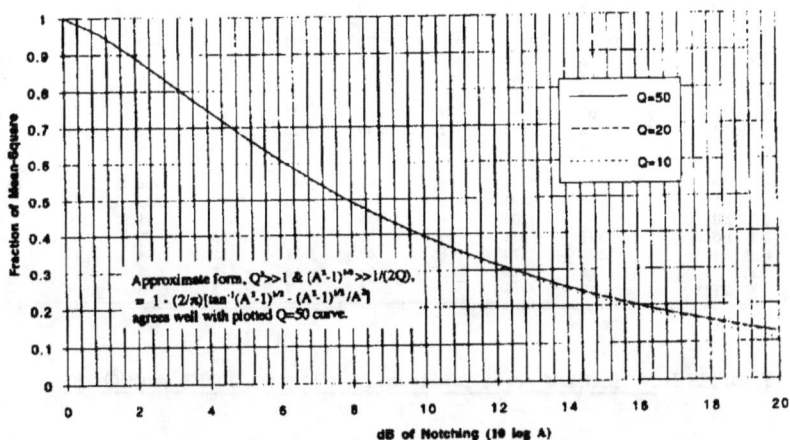

FIGURE 4. Reduction of SDFS Mean-Square Response by Notching

Ratio of modal to residual mass m1/M1, m2/M2	Residual mass ratio, M2/M1								
	0.001	0.003	0.01	0.03	0.1	0.3	1	3	10
8.0, 8.0	932	933	936	948	1001	1180	1240	1234	1238
8.0, 4.0	233	233	233	235	239	256	294	265	250
8.0, 2.0	58	58	58	58	59	60	68	73	68
8.0, 1.0	15	15	15	15	15	15	17	23	22
8.0, 0.5	4	4	4	4	4	4	4	7	6
8.0, 0.25	1	1	1	1	1	1	1	2	5
8.0, 0.125	1	1	1	1	1	1	1	1	3
4.0, 8.0	871	867	858	849	904	1042	1067	1110	1229
4.0, 4.0	218	218	217	216	220	250	254	250	252
4.0, 2.0	55	55	55	55	56	61	72	68	67
4.0, 1.0	14	14	14	14	14	16	21	23	22
4.0, 0.5	3	3	4	4	4	4	6	10	10
4.0, 0.25	1	1	1	1	1	1	2	5	5
4.0, 0.125	1	1	1	1	1	1	1	3	3
2.0, 8.0	1586	1478	1260	1061	990	946	982	1099	1201
2.0, 4.0	406	391	355	305	272	259	238	236	254
2.0, 2.0	103	101	97	88	79	82	70	65	62
2.0, 1.0	26	26	26	25	24	25	25	23	22
2.0, 0.5	7	7	7	7	7	9	10	10	10
2.0, 0.25	2	2	2	2	2	3	5	5	6
2.0, 0.125	1	1	1	1	1	1	3	3	4
1.0, 8.0	11041	5731	2714	1486	967	901	984	1095	1181
1.0, 4.0	3869	2206	1105	567	332	247	233	238	248
1.0, 2.0	1228	826	432	226	125	83	71	66	64
1.0, 1.0	359	283	166	100	50	34	26	23	23
1.0, 0.5	100	89	63	42	24	15	12	11	11
1.0, 0.25	28	27	23	17	11	8	6	6	6
1.0, 0.125	8	8	8	7	5	5	4	4	4
0.5, 8.0	13889	7720	3501	1726	1023	880	974	1093	1171
0.5, 4.0	4516	2895	1417	695	357	247	225	240	244
0.5, 2.0	1346	1003	561	283	136	89	70	64	65
0.5, 1.0	377	319	211	117	59	39	27	24	22
0.5, 0.5	102	95	74	48	27	17	12	11	10
0.5, 0.25	28	27	25	19	13	8	7	6	6
0.5, 0.125	8	8	8	8	6	5	4	4	4
0.25, 8.0	17378	9978	4092	1944	1017	833	936	1092	1166
0.25, 4.0	5194	3725	1805	812	380	249	225	241	242
0.25, 2.0	1455	1205	741	359	173	93	71	66	65
0.25, 1.0	391	354	269	160	74	43	28	23	22
0.25, 0.5	103	99	86	63	38	22	14	12	11
0.25, 0.25	28	28	27	23	16	10	8	7	7
0.25, 0.125	8	8	8	8	7	5	5	4	4

TABLES I. <u>Normalized Force Spectrum for Complex TDFS with Q=20</u>

APPENDICES

Appendix A--Equations for Calculating the Simple TDFS Force Limits

The force limit is calculated for the TDFS in Fig. 1 with different masses of the source and the load oscillators. For this TDFS, the maximum response of the load and therefore the maximum interface force occur when the uncoupled resonance frequency of the load equals that of the source. For this case, the characteristic equation is that of a classical dynamic absorber, from "Vibration-Test Force Limits Derived from Frequency-Shift Method":

$$(\omega/\omega_0)^2 = 1 + (m_2/m_1)/2 \pm [(m_2/m_1) + (m_2/m_1)^2/4)]^{0.5} \tag{A1}$$

where ω_0 is the natural frequency of one of the uncoupled oscillators, m_1 is the mass of the source oscillator, and m_2 is the mass of the load oscillator in Fig. 1. The ratio of the interface force S_{FF} to acceleration S_{AA} spectral densities, divided by the magnitude squared of the load dynamic mass m_2, is:

$$S_{FF}/(S_{AA} m_2^2) = [1 + (\omega/\omega_0)^2/Q_2^2] / \{[1 - (\omega/\omega_0)^2]^2 + (\omega/\omega_0)^2/Q_2^2\} \tag{A2}$$

where Q_2 is the quality factor, one over twice the critical damping ratio, of the load.

The force spectral density, normalized by the load mass squared and by the acceleration spectral density, at the two coupled system resonances is obtained by combining Eqs. (A1) and (A2). For this TDFS the normalized force is just slightly larger at the lower resonance frequency of Eq. (A1). The maximum normalized force spectral density, obtained by evaluating Eq. (A2) at the lower resonance frequency, is plotted against the ratio of load to source mass for three values of Q_2 in Fig. 2.

Appendix B--Calculation of Effective Mass

Applying the rationale of "Equivalent Spring-Mass System: A Physical Interpretation" and subdividing the displacement vector into unrestrained absolute displacements u_f and prescribed absolute displacements u_p, the equilibrium equation is:

$$\begin{bmatrix} m_{FF} & m_{FP} \\ m_{PF} & m_{PP} \end{bmatrix} \begin{Bmatrix} d^2u_F/dt^2 \\ d^2u_P/dt^2 \end{Bmatrix} + \begin{bmatrix} k_{FF} & k_{FP} \\ k_{PF} & k_{PP} \end{bmatrix} \begin{Bmatrix} u_F \\ u_P \end{Bmatrix} = \begin{Bmatrix} f_F \\ f_P \end{Bmatrix} \tag{B1}$$

Let:
$$\{u\} = \phi U = \begin{bmatrix} \phi_N & \phi_P \\ 0 & I_{PP} \end{bmatrix} \begin{Bmatrix} U_N \\ U_P \end{Bmatrix} \tag{B2}$$

Where ϕ_N are normal modes and ϕ_R are rigid body modes associated with a kinematic set of unit prescribed motions, and U_N is the generalized modal relative displacement and U_P is the generalized prescribed absolute displacement. Substituting and pre-multiplying by ϕ^T yields:

$$\left[\frac{[M_{NN} \mid M_{NP}]}{[M^T_{NP} \mid M_{PP}]} \right] \left\{ \frac{d^2 U_N/dt^2}{d^2 U_P/dt^2} \right\} + \left[\frac{[\omega^2_N M_{NN} \mid 0]}{[\quad 0 \quad \mid 0]} \right] \left\{ \frac{U_N}{U_P} \right\} = \left\{ \frac{F_F}{F_P} \right\} \quad (B3)$$

where:

$$M_{NN} = \phi_N^T m_{FF} \phi_N \tag{B4}$$

$$M_{NP} = \phi_N^T m_{FF} \phi_P + \phi_N^T m_{FP} I_{PP} \tag{B5}$$

$$M_{PP} = I_{PP} m_{PP} I_{PP} + I_{PP} m_{PF} \phi_P + \phi_P^T m_{FP} I_{PP} + \phi_P^T m_{FF} \phi_P \tag{B6}$$

$$F_P = I_{PP} f_P \tag{B7}$$

For: $d^2 U_P /dt^2 = U_P = F_F = 0$, $d^2 U_n /dt^2 = -\omega^2 U_n$, and for $U_n = 1$:

$$M_{nP}^T = -F_P / \omega_n^2 \tag{B8}$$

where n indicates a single mode. (Note that M_{nP}^T is in mass units.) M_{nP} is sometimes called the **elastic-rigid coupling** or the **modal participation factor** for the nth mode. If the model is restrained at a single point, the reaction (F_P) in (B8) is the SPCFORCE at that point in a NASTRAN modal analysis.

The initial value of M_{PP} is the rigid body mass matrix. If a Gaussian decomposition of the total modal mass in (B3) is performed, it subtracts the contribution of each normal mode, called the **effective mass**:

$$M_{nP}^T M_{nn}^{-1} M_{nP}, \tag{B9}$$

from M_{PP}^n, which is the **residual mass** after excluding the mass associated with the already processed n modes.

Consider the ratio of the reaction force in a particular direction p, to the prescribed acceleration in a particular direction q; the effective mass, $M_{np}^T M_{nn}^{-1} M_{nq}$, is the same as the contribution of the n th mode to this ratio, divided by the single-degree-of-freedom frequency response factor. The sum of the common-direction effective masses for all modes is equal to the total mass, or moment of inertia for that direction. Values of the effective mass are independent of the modal normalization.

16

Appendix C--Force Specification Example

ds--02/27/97

								FORCE SPECIFICATION EXAMPLE									
						ACE SPACECRAFT		CRIS INSTRUMENT		Y-AXIS (NORMAL TO PANEL)							
Damping (Q = 50, 20, or 5)	Q =	20.00		ζ =	0.025												
Frequency (f)	Hz	40	50	63	80	100	125	160	200	250	315	400	500	630	800	1000
(ω)	rad/sec	251	314	396	503	628	785	1005	1257	1571	1979	2513	3142	3958	5027	6283
Acceleration PSD	[G^2/HZ]	0.01	0.02	0.03	0.04	0.06	0.10	0.12	0.12	0.12	0.12	0.16	0.16	0.16	0.16	0.16
SOURCE Eff Weight [lbs], Total = 488.88 NOTE 1																
Residual Weight (M1)		400.00	400.00	251.95	156.25	100.00	64.00	39.06	25.00	16.00	10.08	6.25	4.00	2.52	1.56	1.00
Total Modal Weight in Band		0.00	0.00	148.05	95.70	56.25	36.00	24.94	14.06	9.00	5.92	3.83	2.25	1.48	0.96	0.56
Number of Modes in Band (N1)		0	0	1	1	1	1	1	1	1	1	1	1	1	1	1
Band Aver. Modal Weight (m1)		0.00	0.00	148.05	95.70	56.25	36.00	24.94	14.06	9.00	5.92	3.83	2.25	1.48	0.96	0.56
Ratio of Modal to Residual Weight (a1)		0.00	0.00	0.59	0.61	0.56	0.56	0.64	0.56	0.56	0.59	0.61	0.56	0.59	0.61	0.56
LOAD Eff Weight [lbs], Total = 66.00 NOTE 1																
Residual Weight (M2)		66.00	66.00	66.00	66.00	66.00	66.00	66.00	52.80	42.24	33.52	26.40	21.12	16.76	13.20	10.56
Total Modal Weight in Band		0.00	0.00	0.00	0.00	0.00	0.00	0.00	13.20	10.56	8.72	7.12	5.28	4.36	3.56	2.64
Number of Modes in Band (N2)		0	0	0	0	0	0	0	1	1	1	1	1	1	1	1
Band Aver. Modal Weight (m2)		0.00	0.00	0.00	0.00	0.00	0.00	0.00	13.20	10.56	8.72	7.12	5.28	4.36	3.56	2.64
Ratio of Modal to Residual Weight (a2)		0.00	0.00	0.00	0.00	0.00	0.00	0.00	0.25	0.25	0.26	0.27	0.25	0.26	0.27	0.25
M2/M1 (Residual)		0.17	0.17	0.26	0.42	0.66	1.03	1.69	2.11	2.64	3.33	4.22	5.28	6.65	8.45	10.56
MAX NORM FORCE (Simple) NOTE 2		9	9	6	4	3	3	2	2	2	2	1	1	1	1	1
MAX NORM FORCE (Complex) NOTE 2		1	1	1	1	1	1	1	6	6	7	7	6	6	6	6
Simple TDFS Force Spec. Note 3 [lb^2/Hz]		389.96	584.94	684.31	777.94	915.19	1119.62	1049.57	609.90	357.80	207.99	159.93	98.50	57.67	34.16	13.18
Complex TDFS Force Spec. Note 3 [lb^2/Hz]		43.56	65.34	108.90	174.24	274.43	435.60	522.72	2007.24	1284.64	891.76	759.41	457.58	276.92	176.12	70.91
Semi-empirical Force Spec. Note 4 [lb^2/Hz]		98.01	147.02	245.03	392.04	617.48	980.10	1176.12	1176.12	752.72	474.12	392.04	250.91	158.04	98.01	39.20

NOTE 1. Residual weight of all higher frequency modes.
NOTE 2. Normalized by load residual mass squared and by acceleration spectrum.
NOTE 3. See "Vibration-Test Force Limits Derived from Frequency-Shift Method", AIAA Jour. Spacecraft and Rockets, pp. 312-316, March-April 1995.
NOTE 4. See "Cassini Spacecraft and Instrument Force Limited Vibration Testing", ESTEC, Noordwijk, The Netherlands, 24-27 June 1997, with C=1.5

17

REPORT DOCUMENTATION PAGE

Public reporting burden for this collection of information is estimated to average 1 hour per response, including the time for reviewing instructions, searching existing data source, gathering and maintaining the data needed, and completing and reviewing the collection of information. Send comments regarding this burden estimate or any other aspect of collection of information, including suggestions for reducing this burden, to Washington Headquarters Services, Directorate for Information Operations and Reports, 1215 Jeffers Davis Highway, Suite 1204, Arlington, VA 22202-4302, and to the Office of Management and Budget, Paperwork Reduction Project (0704-0188), Washington, DC 20503.

1. AGENCY USE ONLY *(Leave blank)*	2. REPORT DATE May 1997	3. REPORT TYPE AND DATES COVERED Final

4. TITLE AND SUBTITLE

Force Limited Vibration Testing Monograph

5. FUNDING NUMBERS

C - NAS7-1260

RF 295 Am2

6. AUTHOR(S)

Terry D. Scharton, Ph.D.

7. PERFORMING ORGANIZATION NAME(S) AND ADDRESS(ES)

Jet Propulsion Laboratory
California Institute of Technology
4800 Oak Grove Drive
Pasadena, CA 91109-8099

8. PERFORMING ORGANIZATION REPORT NUMBER

9. SPONSORING / MONITORING AGENCY NAME(S) AND ADDRESS(ES)

National Aeronautics and Space Administration
Washington, DC 20546-0001

10. SPONSORING / MONITORING AGENCY REPORT NUMBER

11. SUPPLEMENTARY NOTES

12a. DISTRIBUTION / AVAILABILITY STATEMENT

Public Distribution   Subject Category 39

12b. DISTRIBUTION CODE

13. ABSTRACT *(Maximum 200 words)*

The practice of limiting the shaker force in vibration tests was instigated at the NASA Jet Propulsion Laboratory (JPL) in 1990 after the mechanical failure of an aerospace component during a vibration test. Now force limiting is used in almost every major vibration test at JPL and in many vibration tests at NASA Goddard Space Flight Center (GSFC) and at many aerospace contractors. The basic ideas behind force limiting have been in the literature for several decades, but the piezo-electric force transducers necessary to conveniently implement force limiting have been available only in the last decade. In 1993, funding was obtained from the NASA headquarters Office of Chief Engineer to develop and document the technology needed to establish force limited vibration testing as a standard approach available to all NASA centers and aerospace contractors. This monograph is the final report on that effort and discusses the history, theory, and applications of the method in some detail.

14. SUBJECT TERMS

Force limiting; Vibration; Vibration testing; Impedance; Random vibration; Response limiting; Force transducers; Notching

15. NUMBER OF PAGES

114

16. PRICE CODE

17. SECURITY CLASSIFICATION OF REPORT	18. SECURITY CLASSIFICATION OF THIS PAGE	19. SECURITY CLASSIFICATION OF ABSTRACT	20. LIMITATION OF ABSTRACT
Unclassified	Unclassified	Unclassified	Unlimited

www.ingramcontent.com/pod-product-compliance
Lightning Source LLC
Chambersburg PA
CBHW080301180526
45167CB00006B/2628